Quantitative Applications of Mass Spectrometry

Quantitative Applications of Mass Spectrometry

Irma Lavagnini
University of Padova, Padova, Italy

Franco Magno
University of Padova, Padova, Italy

Roberta Seraglia
National Council of Research, Padova, Italy

Pietro Traldi
National Council of Research, Padova, Italy

JOHN WILEY & SONS, LTD

Telephone (+44) 1243 779777

Email (for orders and customer service enquiries): cs-books@wiley.co.uk
Visit our Home Page on www.wileyeurope.com or www.wiley.com

Other Wiley Editorial Offices

John Wiley & Sons Inc., 111 River Street, Hoboken, NJ 07030, USA

Jossey-Bass, 989 Market Street, San Francisco, CA 94103-1741, USA

Wiley-VCH Verlag GmbH, Boschstr. 12, D-69469 Weinheim, Germany

John Wiley & Sons Australia Ltd, 42 McDougall Street, Milton, Queensland 4064, Australia

John Wiley & Sons (Asia) Pte Ltd, 2 Clementi Loop #02-01, Jin Xing Distripark, Singapore 129809

John Wiley & Sons Canada Ltd, 22 Worcester Road, Etobicoke, Ontario, Canada M9W 1L1

Wiley also publishes its books in a variety of electronic formats. Some content that appears in print may not be available in electronic books.

Library of Congress Cataloging-in-Publication Data

Quantitative applications of mass spectrometry/Irma Lavagnini ... [et al.].
p. cm.
Includes bibliographical references and index.
ISBN-13: 978-0-470-02516-1 (pbk. : acid-free paper)
ISBN-10: 0-470-02516-6 (pbk. : acid-free paper)
1. Mass spectrometry. 2. Chemistry, Analytic–Quantitative.
I. Lavagnini, Irma.
QD96.M3Q83 2006
543′.65–dc22 2005036663

British Library Cataloguing in Publication Data

A catalogue record for this book is available from the British Library

ISBN-13 978-0-470-02516-1 (Paperback)
ISBN-10 0-470-02516-6 (Paperback)

Typeset in 10/12 pt Times by Thomson Press (India) Ltd, New Delhi, India
Printed and bound in Great Britain by TJ International Ltd, Padstow, Cornwall
This book is printed on acid-free paper responsibly manufactured from sustainable forestry in which at least two trees are planted for each one used for paper production.

To our past, present and future students who stimulate our interest in the research and who, hopefully, have learnt or will learn something from our efforts

Contents

Preface

This book has been born from the long-term collaboration (and friendship!) existing between two research groups operating in the Padova area. The first has operated for more than 30 years in the field of analytical chemistry, the second for nearly the same amount of time in organic mass spectrometry. The exchange of specific knowledge and experiences between the two groups has been very fruitful, in particular in the development phase of quantitative analyses by mass spectrometry. Both operative and theoretical aspects have been the objects of many discussions and this was fundamental to clarify those doubts that have arisen for those working in the research and analytical fields.

In the last two decades mass spectrometry has shown a phenomenal growth and nowadays it is an essential tool in environmental and biomedical fields. The problem that can arise from this wide expansion is that mass spectrometry is often mainly considered as a 'magic box' in which on one side a sample is introduced and on the other side the analytical data come out. The software (and the marketing!) has removed all the doubts and critical analysis of the data.

With this book we wish to present some very basic information to the scientists and technicians working in the field of quantitative organic mass spectrometry.

Our efforts have been devoted to authoring a book that is easy to read for researchers who are not necessarily physicists or chemists, but mainly for those who, for the first time, face all the problems arising from the development and use of a quantitative procedure. The last chapter presents a description of the theoretical aspects related to calibration and data analysis and is devoted to those who wish to learn more about these aspects.

The picture we have chosen for the cover is a Dolomite peak. Leaving aside the banal shape similarity with a chromatographic peak, this choice was made because the mountain is a good teacher of life: if you want to reach the top of the peak directly you must exert a lot of energy or, alternatively, you must study its structure and choose the right way to reach the summit with less effort. In other words each mountain climbed requires both a general strategy and many tactical choices to be performed along the way. Thus, the same approach must be used in the development of a quantitative analysis by mass spectrometry: be sure of each step you are doing, otherwise the peak will remain out of reach!

Acknowledgements

The authors wish to thank sincerely Dr. Roberta Zaugrando (Venice University) for the dioxine analysis data and Profs. Gloriano Moneti and Giuseppe Pieraccini (Florence University) for the data related to testosterone analysis.

Introduction

Nowadays, mass spectrometry (MS) is one of the most frequently employed techniques in performing quantitative analysis. Its specificity, selectivity and typical limit of detection are more than enough to deal with most analytical problems.

This is the result of significant effort, either from the scientists working in the field or from the manufacturing industry, devoted to the development of new ionization methods, expanding the application fields of the technique, and new analysers capable of increasing the specificity mainly by collisional experiments (MS/MS or "tandem mass spectrometry") or by high mass accuracy measurements.

Thus, the MS panorama is made up of many instrumental configurations, each of which have specific positive and negative aspects and different cost/benefit ratios.

Of course, these mass spectrometric approaches are usually employed when linked to suitable chromatographic (C) systems. The synergism obtained allows C-MS to be used worldwide and is of considerable interest to researchers involved in basic chemistry, environmental and food controls, biochemistry, biology and medicine.

It is to be expected that this diffusion will grow in the future, due to the relevance of the information that quantitative MS can provide, in particular in the field of public health. For this reason, some basic information on the phenomena which form the basis of different instrumental approaches, the general strategy to be employed for the development of a quantitative analysis, the role of the specificity in this context and some theoretical aspects on calibration and data analysis, are of interest and this book aims to cover, as simply as possible, all these aspects.

1

What Instrumental Approaches are Available

The fantastic development of mass spectrometry (MS) in the last 30 years has led this technique to be applied practically in all analytical fields. We focus our attention on the application in the organic, biological and medical fields which nowadays represent the environment in which MS finds the widest application. This chapter is devoted to a short description of the different instrumental approaches currently in use and commercially available.

MS is based on the production of ions from the analyte, their analysis with respect to their mass to charge ratio (m/z) values and their detection. Consequently, at instrumental level three components are essential to perform mass spectrometric experiment: (i) ion source; (ii) mass analyser; and (iii) detector (Figure 1.1). Of course, the performances of these three components reflect on the quality of both quantitative and qualitative data. It must be emphasized that generally these three components are spatially separated (Figure 1.1a) and only in two cases [Paul ion trap and Fourier transform mass spectrometer without external source(s)] can they occupy the same physical space and, consequently, the ionization and mass analysis must be separated in time (Figure 1.1b).

1.1 ION SOURCES

The ion production is the phenomenon which highly affects the quality of the mass spectrometric data obtained. The choice of the ionization

Quantitative Applications of Mass Spectrometry I. Lavagnini, F. Magno, R. Seraglia and P. Traldi

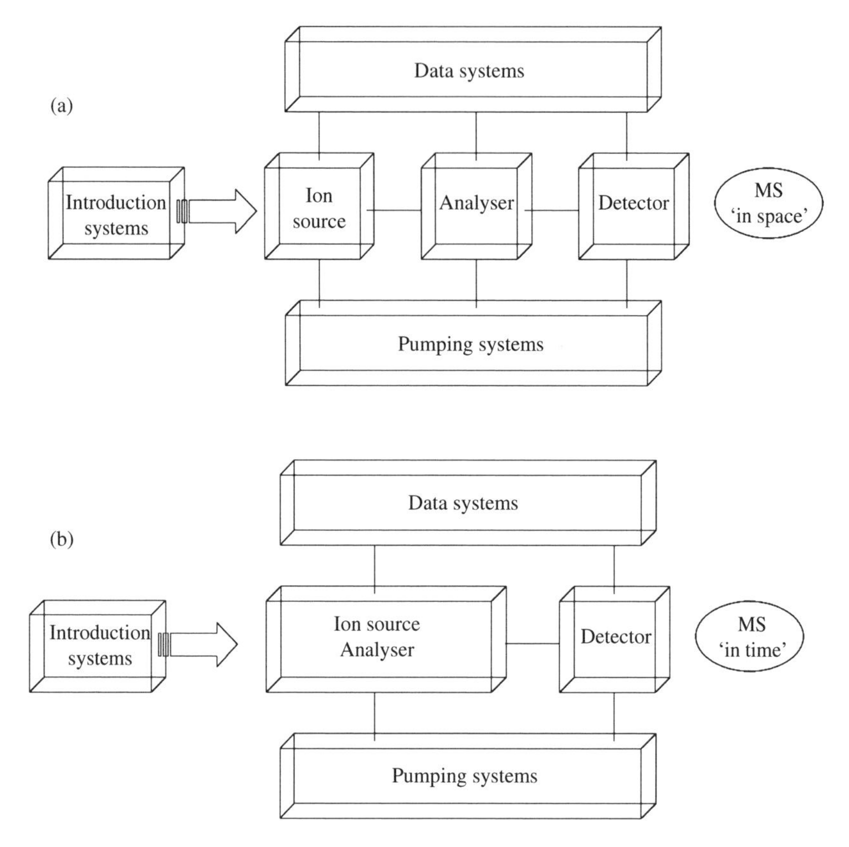

Figure 1.1 Schemes of MS systems 'in space' (a) and 'in time' (b)

method to be employed is addressed by the physico-chemical properties of the analyte(s) of interest (volatility, molecular weight, thermolability, complexity of the matrix in which the analyte is contained).

Actually the ion sources usually employed can be subdivided into two main classes: those requiring sample in the gas phase prior to ionization; and those able to manage low volatility and high molecular weight samples.

The first class includes electron ionization (EI) and chemical ionization (CI) sources which represent those worldwide most diffused, due to their extensive use in GC/MS systems. The other ones can be further divided into those operating with sample solutions [electrospray ionization (ESI), atmospheric pressure chemical ionization (APCI), atmospheric pressure photoionization (APPI)] and those based on the contemporary sample desorption and ionization from a solid substrate [matrix - assisted laser desorption/ionization (MALDI) and LDI].

1.1.1 Electron Ionization

EI is based on the interaction of an energetic (70 eV) electron beam with the sample vapour (at a pressure in the range 10^{-7}–10^{-5} Torr) (Figure 1.2). This interaction leads to the production of a series of ions related to the chemical properties of the compound(s) under study. The theoretical treatment of EI is beyond the scope of the present book and it is possible to find it in many publications.[1] For the present discussion it is enough to consider that EI generally leads to a molecular ion $M^{+\bullet}$, originating by the loss of an electron from the neutral molecule:

$$M + e^- \rightarrow M^{+\bullet} + 2e^-$$

and to a series of fragments, generally highly diagnostic from the structural point of view:

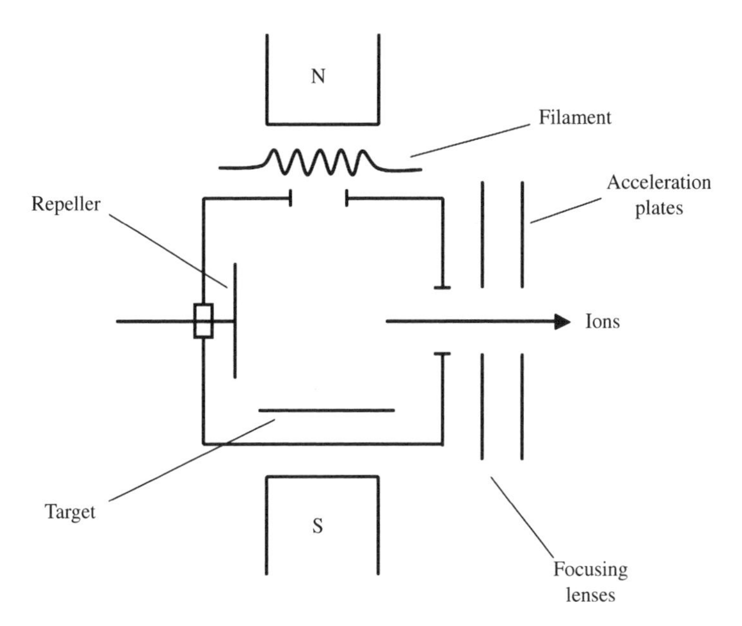

Figure 1.2 Scheme of an electron ionization (EI) ion source. The electrons are generated by the filament, accelerated by the potential between the filament and ion chamber, and focused on the target. A permanent magnet (N–S) is mounted axially to the electron beam to induce a cycloidal pathway (and a consequent increase of electron-neutral interactions), leading to higher ion production yield. The repeller electrode favours the acceleration field penetration, leading to higher ion extraction

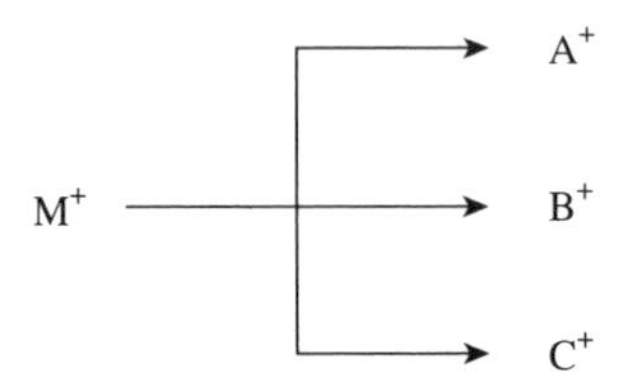

Some of them originate from simple bond cleavages, while some others are produced through rearrangement processes. What must be emphasized is that EI leads to well reproducible mass spectra. In other words, by different EI sources spectra practically superimposable are obtained and this is the reason for which the only spectrum libraries available are those based on EI data.

The main efforts done in the last decade in the EI field are due to the development of ion sources with the highest possible ion yield. To reach this aim, on the one hand an optimization of ion source geometry has been performed (this has been achieved by the development of suitable ion optics to increase either the ion production or the ion extraction), on the other, to make inert the ion source walls (originally in stainless steel) so as to avoid the sample loss due to its pyrolysis on the hot metallic surface.

The quantitative data obtained by EI can be strongly affected mainly by two parameters: the first related to sample loss (due to problems related to sample injection lines and to 'open' source configuration as well as to thermal decompositions occurring in injection lines and/or source), while the second can be related to a decreased efficiency of ion extraction (nonoptimized extraction field, field modification due to the presence of polluted surfaces). These two aspects reflect not only on the limit of detection (LOD) of the system but also on the linearity of the quantitative response.

The ion most diagnostic from the qualitative point of view is usually considered the molecular one ($M^{+\bullet}$). However, wide classes of compounds, easily vaporized, do not lead to the production of $M^{+\bullet}$. This is due to the energetics of EI induced decomposition processes. In other words if a decomposition process is energetically favoured (with a particularly low critical energy) it takes place immediately, due to the internal energy content of $M^{+\bullet}$. To overcome this problem in the 1960s a new ionization method was developed, based on gas-phase chemical reactions.

1.1.2 Chemical Ionization

To obtain a lower energy deposition in the molecule of interest, reflecting in the privileged formation of charged molecular species, in the 1960s CI methods were proposed.[2] They are based on the production in the gas phase of acidic or basic species, which further react with a neutral molecule of analyte leading to $[M+H]^+$ or $[M-H]^-$ ions, respectively. Generally, protonation reactions of the analyte are those more widely employed; the occurrence of such reactions is related to the proton affinity (PA) of M and the reactant gas, and the internal energy of the obtained species are related to the difference between these proton affinities. Thus, as an example, considering an experiment performed on an organic molecule with PA value of 180 kcal/mol (PA_M), it can be protonated by reaction with CH_5^+ ($PA_{CH_4} = 127$ kcal/mol), H_3O^+ ($PA_{H_2O} = 165$ kcal/mol), but not with NH_4^+ ($PA_{NH_3} = 205$ kcal/mol).[3] This example shows an important point about CI: it can be effectively employed to select species of interest in complex matrices. In other words, by a suitable selection of a reacting ion $[AH]^+$ one could produce $[MH]^+$ species of molecules with PA higher than that of A. Furthermore the extension of fragmentation can be modified in terms of the difference of $[PA_M - PA_A]$.

From the operative point of view CI is simply obtained by introducing the neutral reactant species inside an EI ion source in a 'close' configuration, by which quite high reactant pressure can be obtained (Figure 1.3).

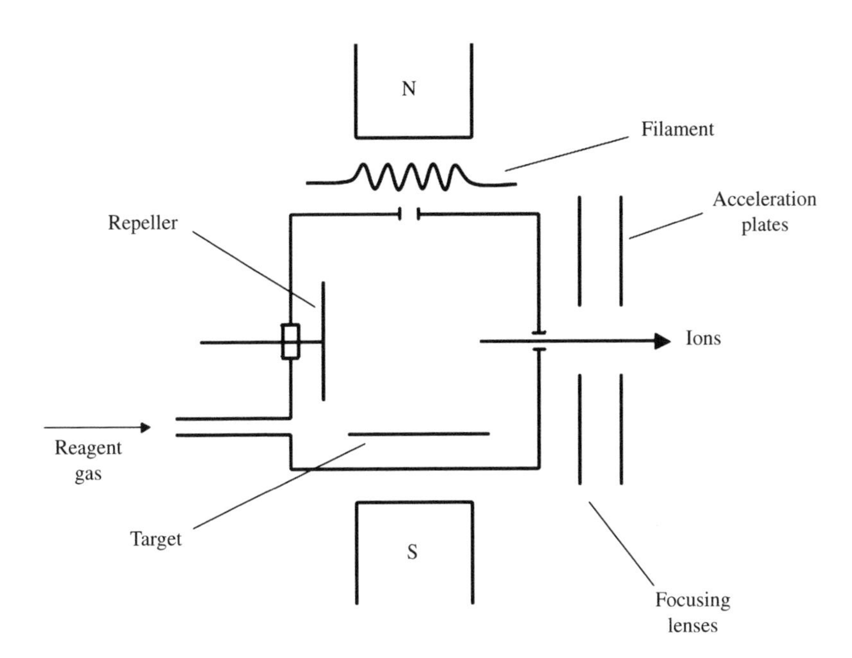

Figure 1.3 Scheme of a chemical ionization (CI) ion source. The electron entrance and ion exit holes are of reduced dimension in order to obtain, inside the ion chamber, effective pressure of the reagent gas

If the operative conditions are properly set the formation of abundant $[AH]^+$ species (or, in the case of negative ions B^-) is observed in high yield. Of course, attention must be paid in particular in the case of quantitative analysis to reproduce carefully these experimental conditions, because they reflect substantially on the LOD values.

CI, as well as EI, requires the presence of samples in vapour phase and consequently it cannot be applied for nonvolatile analytes. Efforts have been made from the 1960s to develop ionization methods overcoming these aspects and, among them, field desorption (FD)[4] and fast atom bombardment (FAB)[5] resulted in highly effective methods and opened new applications for mass spectrometry. More recently new techniques have become available and are currently employed for nonvolatile samples: APCI,[6] ESI,[7] APPI[8] and MALDI[9] represent nowadays the most used for the analysis of high molecular weight, high polarity samples.

For these reasons, we describe these methods.

1.1.3 Atmospheric Pressure Chemical Ionization

APCI[6] was developed starting from the consideration that the yield of a gas-phase reaction does not depend only on the partial pressure of the two reactants, but also on the total pressure of the reaction environment. For this reason the passage from the operative pressure of 0.1–1 Torr, present inside a classical CI source, to atmospheric pressure would, in principle, lead to a relevant increase in ion production and, consequently, to a relevant sensitivity increase.

At the beginning of the research devoted to the development of the APCI method, the problem was the choice of the ionizing device. The most suitable and effective one was, and still is, a corona discharge. The important role of this ionization method mainly lies in its possible application to the analysis of compounds of interest dissolved in suitable solvents: the solution is injected in a heated capillary (typical temperatures in the range 350–400 °C), which behaves as a vaporizer. The solution is vaporized and reaches outside from the capillary the atmospheric pressure region where the corona discharge takes place. Usually the vaporization is assisted by a nitrogen flow coaxial to the capillary (Figure 1.4). The ionization mechanism is typically the same present in CI experiments (Figure 1.5). The solvent molecules, present in high abundance, are statistically privileged to interact with the electron beam originated from the corona discharge; the ions so formed react with other solvent molecules leading to

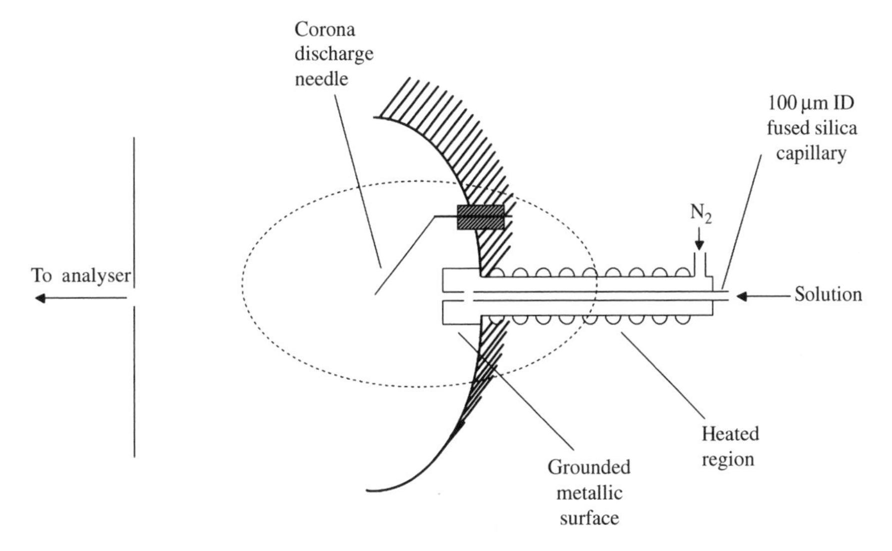

Figure 1.4 Scheme of an APCI ion source

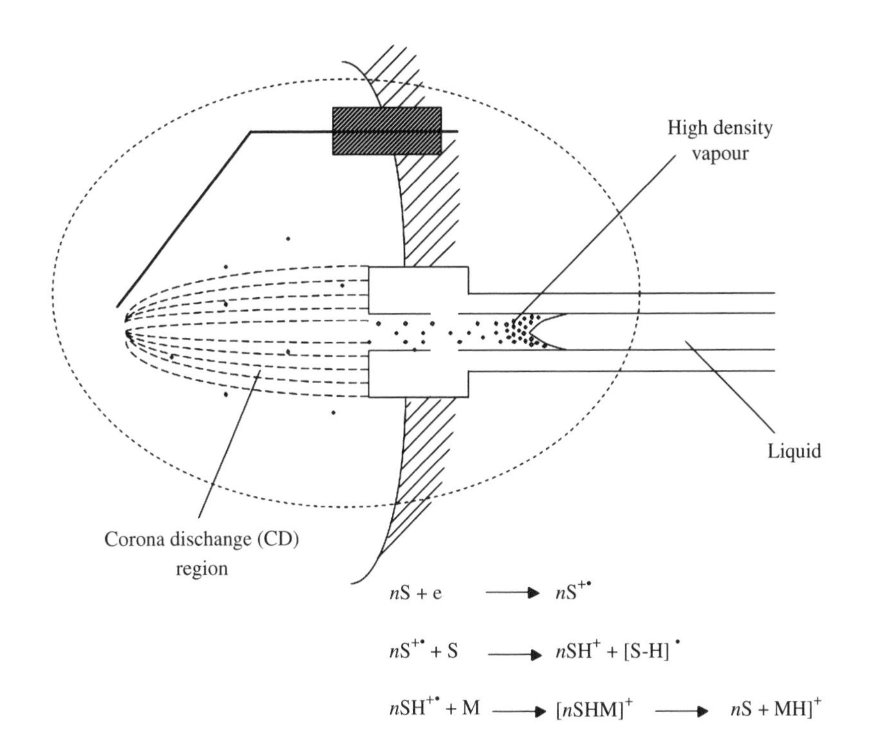

Figure 1.5 Corona discharge region of an APCI source and reactions occurring in it

protonated (in the case of positive ions analysis) or deprotonated (negative ions analysis) species, which are the reactant for the analyte ionization. One problem which, at the beginning of its development, APCI exhibited was the presence of analyte molecules still solvated, i.e. the presence of clusters of analyte molecules with different numbers of solvent molecules. To obtain a declustering of these species, different approaches have been proposed, among which nonreactive collision with target gases (usually nitrogen) and thermal treatments are those considered most effective and currently employed. Different instrumental configurations, based on a different angle between the vaporizer and entrance capillary (or skimmer) have been proposed; 180 ° (in line) and 90 ° (orthogonal) geometries are those most widely employed.

In particular, in the case of quantitative analysis, a particular care must be devoted to finding the best operating conditions (vaporizing temperature and solution flow) of the APCI source, which lead to the most stable signals, and carefully maintaining these conditions for all the measurements.

1.1.4 Electrospray Ionization

ESI[7] is obtained by injection, through a metal capillary line, of solutions of analyte in the presence of a strong electrical field. The production of ions by ESI can be considered as due to three main steps: (i) production of charged drops in the region close to the metal capillary exit; (ii) fast decreasing of the charged drop dimensions due to solvent evaporation and, through phenomena of coulombic repulsion, formation of charged drop of reduced dimension; (iii) production of ions in the gas phase originated from small charged droplets.

The experimental device for an ESI experiment is shown in Figure 1.6. The analyte solution exits from the metal capillary (external diameter, r_c, in the order of 10^{-4} m) to which a potential (V_c) of 2–5 kV is applied; the counter electrode is placed at a distance (d) ranging from 1 to 3 cm. This counter electrode in an ESI source is usually a skimmer with a 10 μ orifice or an 'entrance heated capillary' (internal diameter 100–500 μ; length 5–10 cm), which represents the interface to the mass spectrometric analyser. Considering the thickness of the metal capillary, the electrical field (E_c) close to it is particularly high. For example, for $V_c = 2000$ V, $r_c = 10^{-4}$ m and $d = 0.02$ m, an E_c in the order of 6×10^6 V/cm has been calculated by Pfeifer and Hendricks.[10] This

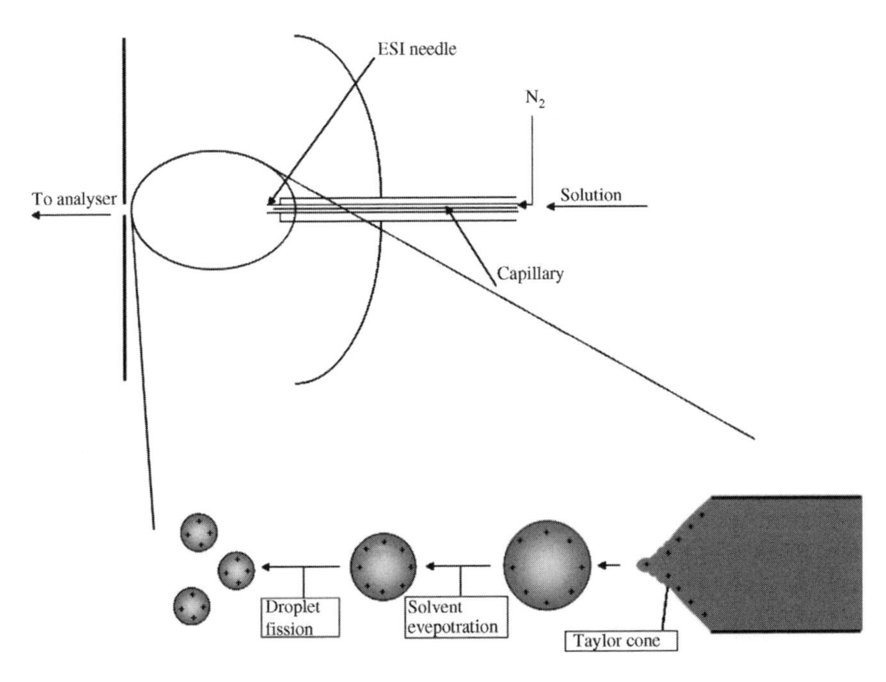

Figure 1.6 Scheme of ESI ion source and enlarged view of droplet generation region. (Taylor cone and droplet dimension are not to the same scale)

electrical field interacts with solution and the charged species present inside the solution move in the field direction, leading to the formation of the so called 'Taylor cone'.[11] If the electrical field is high enough, a spray is formed from the cone apex, consisting of small charged droplets. In the case of positive ion analysis, i.e. when the needle is placed at a positive voltage, the droplets bear positive charge and vice versa in the case of negative ion analysis. A charged drop moves through the atmosphere for the field action in the direction of the counter electrode. The solvent evaporation leads to the reduction of the drop dimensions and to a consequent increase of the electrical field perpendicular to the droplet surface. For a specific value of droplet radius the ion repulsion becomes stronger than surface tension and in these conditions the droplet explosion takes place.

Two mechanisms have been proposed for the formation of gaseous ions from small charged droplets. The first model, called 'charge residue mechanism' (CRM) was proposed by Dole in 1968[12] and describes the process as sequential scissions leading to the production of small droplets bearing one or more charges but only one analyte molecule. When the last, few solvent molecules evaporate the charge(s) remains deposited on the analyte structure, which gives rise to the most stable gaseous ion.

More recently, Iribarne and Thomson have proposed a different mechanism, describing the direct emission of gaseous ions from the droplets, after it has reached a certain dimension.[13] This process, called the 'ion evaporation mechanism' (IEM) is predominant on the coulombic fission for droplets of radius, r, lower than 10 μ.

From the above, the reader can consider the factors which can affect the ion production and consequently the sensitivity and reproducibility in ESI measurements. The ion intensity exhibits with respect to analyte concentration a typical trend, analogous to that reported, as an example, in Figure 1.7. A linear portion with a slope of about 1 is present for low concentration until 10^{-6} M, followed by a slow saturation with a weak intensity decreasing at the highest concentrations (10^{-3} M). The linear portion, where intensity is proportional to concentration, is the only region suitable for quantitative analysis. The general trend of the plot can be explained considering that in the system there is not just a single analyte: further electrolytes are always present, for example, impurities, co-analytes and buffer. It should be emphasized that for analyte concentrations lower than 10^{-5} M, the electrospray phenomenon occur due to the presence of electrolytes as impurities, which lead to the electrical conductivity necessary for the 'Taylor cone' production.

Also, in the case of ESI sources, 'in line' or 'orthogonal' geometries have been proposed and employed.

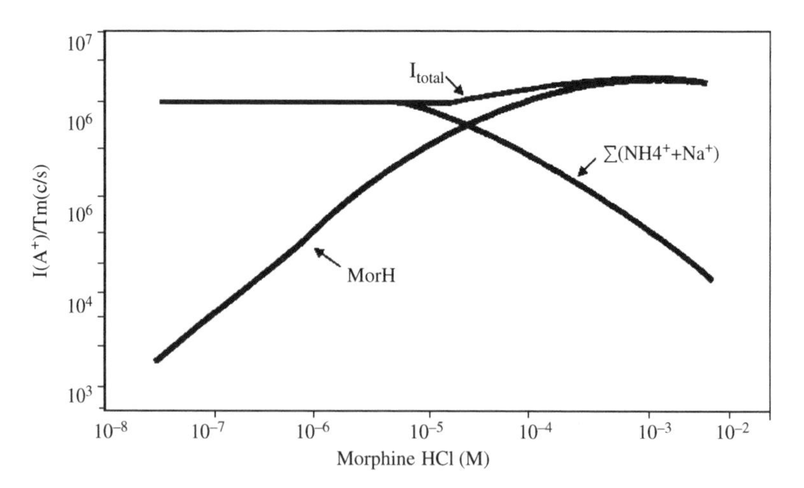

Figure 1.7 Plot of ion intensity *vs* analyte (morphine HCl) concentration obtained by ESI experiments. Reprinted from P. Kebarle and L. Tang, Anal. Chem. 65, 980A (1993), with permission from the American Chemical Society

1.1.5 Atmospheric Pressure Photoionization

A method recently developed consists in the irradiation, by a normal krypton (Kr) lamp, of the vaporized solution of the sample of interest at atmospheric pressure (APPI).[8] The instrumental set up is very similar to that already described for the APCI system (Figure 1.8). In this case, the needle for corona discharge is no longer present, while the solution vaporizer is exactly the same as for the APCI source. On a side of the source the Kr lamp is mounted, so that the vapour solution can be irradiated by photons with energies up to 10.6 eV. The photoionization follows a simple general rule: a molecule with ionization energy (IE_M) can be ionized by photons with energy $E_v = hv$ only when:

$$IE_M \leq E_v$$

Considering that the most of solvents employed in liquid chromatography (LC) methods have an IE higher than 10.6 eV and consequently cannot be ionized by interaction with photon coming from the Kr lamp, the APPI method seems to be, in principle, highly effective for liquid chromatography/mass spectrometry (LC/MS) analysis of compounds

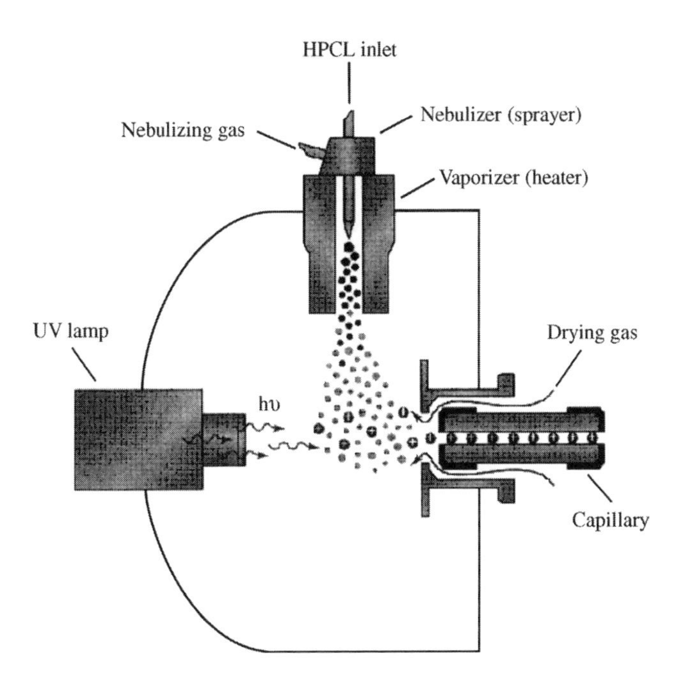

Figure 1.8 Scheme of an APPI ion source

exhibiting IE lower than 10.6 eV. In the case of compounds of interest with IE > 10.6 eV, the use of dopants (i.e. substances photoionizable acting as intermediates in the ionization of the molecule of interest) has been proposed.[8]

Some investigations have shown that some unexpected reactions can take place in the APPI source, indicating that it can be applied not only for analytical purposes but also for fundamental studies of organic and environmental chemistry.[14]

1.1.6 Matrix-assisted Laser Desorption/Ionisation

MALDI[9] consists of the interaction of a laser beam with a solid sample constituted by a suitable matrix in which the analyte is present at very low molar ratio (1:10 000) (Figure 1.9). This interaction leads to the

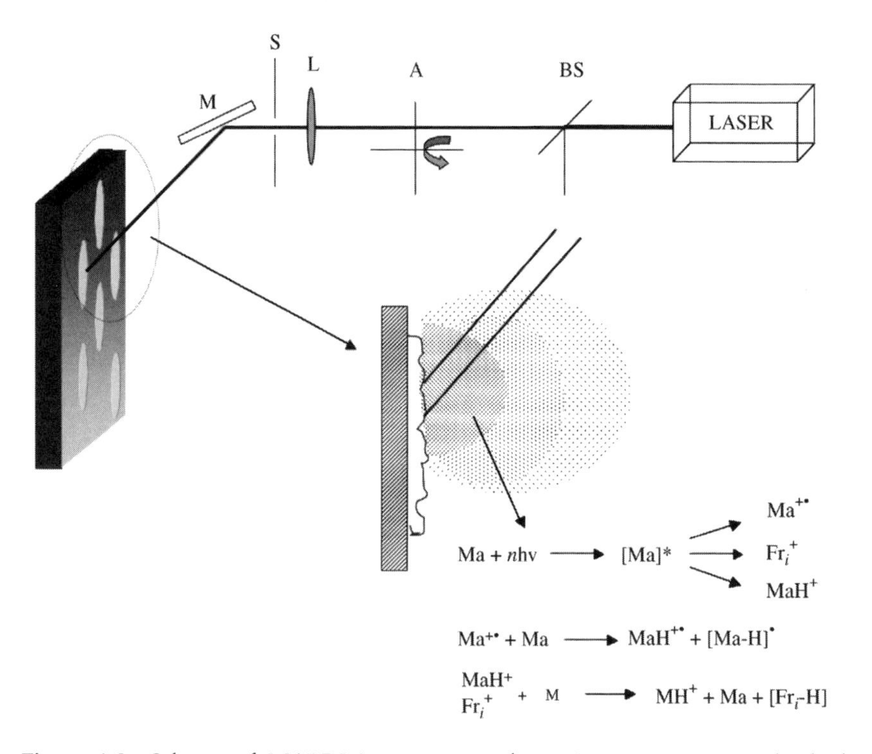

Figure 1.9 Scheme of MALDI ion source and reactions occurring in the high density plume originated by laser irradiation. BS, beam splitter (a portion of the laser beam is used to start the spectrum acquisition); A, attenuator (to regulate the laser beam intensity); L, focusing lens, S, slit; M, mirror

vaporization of a small volume of the solid sample: in the plume of the high density vapour so generated, reactive species originating for the matrix irradiation react with the neutral molecules of analyte, mainly through protonation/deprotonation mechanisms.

A detailed description of the MALDI mechanism is highly complex, due to the presence of many different phenomena:

(i) First of all the choice of the matrix is relevant to obtain effective and well reproducible data.

(ii) The solid sample preparation is usually achieved by the deposition on a metallic surface of the solution of matrix and analyte with concentration suitable to obtain the desired analyte/matrix ratio. The solution is left to dry under different conditions (simply at atmospheric pressure, reduced pressure or under nitrogen stream); in all cases what is observed is the formation of a inhomogeneous solid sample, due to the different crystallization rate of the matrix and analyte. Consequently, the 1:10 000 ratio is only a theoretical datum: in the solid sample different ratios will be found in different positions and the only way to overcome this is to average a high number of spectra corresponding to laser irradiation of different points.

(iii) The photon–phonon transformation, obtained when a photon interacts with a crystal and giving information on the vibrational levels of the crystal lattice, cannot be applied in the laser induced vaporization observed in MALDI experiments, due to the inhomogeneity of the solid sample.

(iv) The laser irradiance is an important parameter: different irradiance values lead to vapour cloud of different density and consequently different ion–molecule reactions can take place.

In other words the MALDI data originate from a series of physical phenomena and chemical interactions originating by the parameterization (matrix nature, analyte nature, matrix/analyte molar ratio, laser irradiation value, averaging of different single spectra), which must be kept under control as much as possible. However, the results obtained by MALDI are of high interest, due to its applicability in fields not covered by other ionization methods. Due to the pulsed nature of ionization phenomena (an N_2 laser operating with pulses of 10^2 ns and with a repetition rate of 5 MHz) the analyser usually employed to obtain the MALDI spectrum is the time-of-flight (TOF) one, which will be described in Section 1.2.4.

1.2 MASS ANALYSERS

The mass analysis of ions in the gas phase is based on their interaction with electrical and magnetic fields. Originally the main component of these devices was a magnetic sector which separates the ions with respect to their *m/z* ratio. Until the 1960s most of the mass spectrometers devoted to physics, organic and organometallic chemistry were based on this approach, and high resolution conditions were (and still are) generally acquired by the use of an electrostatic sector. The double-focusing instruments were (and are) of large dimension (at least 2 m^2) and required the use of heavy magnet and large pumping systems.

In the 1960s, mainly due to the efforts of the Paul group at Bonn University, the development of devices based on electrodynamic fields for mass analysis led to the production of quadrupole mass filters and ion traps of small dimension, so that the mass spectrometer became a bench-top instrument.

The ease of use of these devices, the ease of interfacing them with data systems and, over all, the relatively low cost were the factors that moved mass spectrometry from high level, academic environments to application laboratories, in which the instrument is considered just in terms of its analytical performances.

In this section, the analysers currently most widely employed will be described, in terms of the physical phenomena on which they are based, of their performances and their ease of use.

1.2.1 Mass Resolution

The main characteristic of a mass analyser is its resolution, defined as its capability to separate two neighbouring ions. The resolution necessary to separate two ions of mass M and $(M + \Delta M)$ is defined as:

$$\mathrm{R} = M/\Delta M$$

Then, as an example, the resolution necessary to separate N_2^+ (exact mass = 28.006158) from CO^+ (exact mass = 27.994915) is:

$$\mathrm{R} = M/\Delta M = 28/0.011241 = 2490$$

From the theoretical point of view, the resolution parameters can be described by Figure 1.10. It follows that a relevant parameter is the

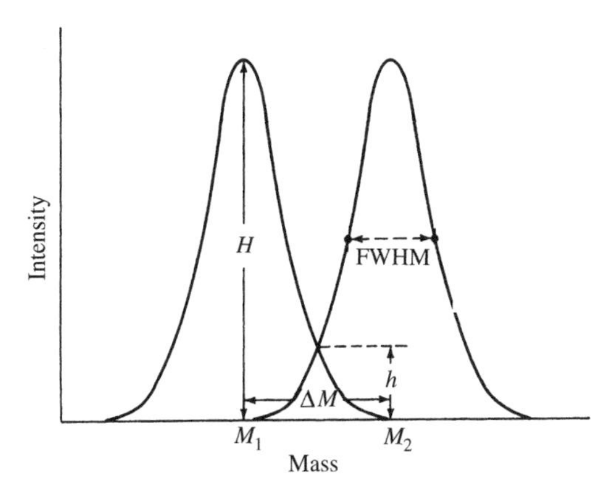

Figure 1.10 Mass resolution parameters

valley existing between the two peaks. Usually resolution data are related to 10 % valley definition.

If the peak shape is approximately gaussian the resolution can be obtained by a single peak. In fact, as shown by Figure 1.10, the mass difference, ΔM, is equal to the peak width at 5 % of its height and, accordingly to the gaussian definition, it is about two times the full width at half maximum height (FWHM). Consequently, by this approach it is possible to estimate the resolution of a mass analyser simply by looking at a single peak, without introduction of two isobaric species of different accurate mass.

The resolution present in different mass analysers can be affected by different parameters and different definitions can be employed. Thus, in the case of a magnetic sector instrument the above 10 % valley definition is usually employed, while in the case of a quadrupole mass filter the operating conditions are such to keep ΔM constant through the entire mass range. Consequently, in the case of a quadrupole mass filter the resolution will be 1000 at *m/z* 1000 and 100 at *m/z* 100, while in the case of magnetic sector the resolution will be, for example, 1000 at *m/z* 1000 and 10 000 at *m/z* 100.

This parameter will be useful to evaluate and compare the performances of different instrumental approaches.

1.2.2 Sector Analysers

At the beginning of the last century, after fundamental studies by Thompson and Aston, which led to the development of the first effective mass spectrograph, the first sector instrument was developed by Dempster.[15] As shown in Figure 1.11, in the Dempster instrument the ions are accelerated by means of a negative potential which could be changed from 500 to 1750 V. The ion beam collimated by the slit S_1 enters in uniform magnetic field B.

From the equations:

$$mv^2/2 = zV(\text{kinetic energy} = \text{potential energy}) \qquad (1.1)$$

and

$$mv^2/R = zvB(\text{centrifugal force} = \text{centripetal force}) \qquad (1.2)$$

it follows that:

$$m/z = B^2R^2/2V \qquad (1.3)$$

where m is ion mass, z is ion charge, V is acceleration potential, v is speed acquired by the ion after acceleration and R is the circular pathway radius of the ion inside the magnetic field.

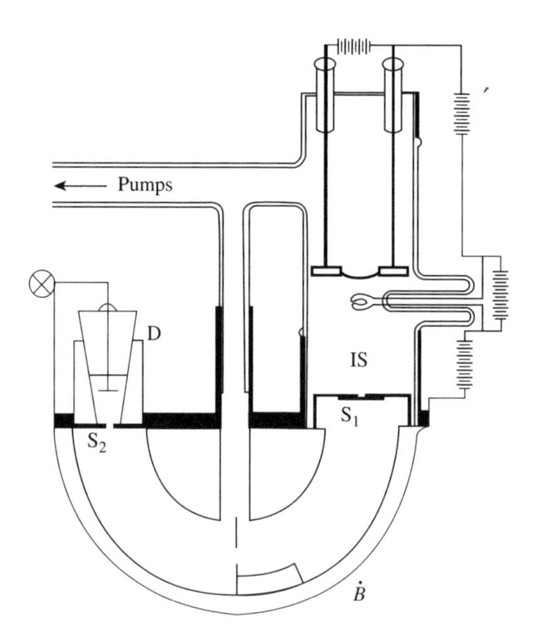

Figure 1.11 Dempster mass spectrometer. IS, ion source; S_1, ion source slit; B, 180° magnetic field; S_2, collector slit; D, electrostatic detector

This shows the capability of a magnetic analyser to separate ions with different m/z ratios with respect to B or V. By scanning B or V it is then possible to focus, through the slit S_2, all the ionic species generated inside the source, separated and ordered with respect to their m/z ratio. Dempster chose to perform the acceleration voltage scan, being difficult at that time to perform regular and reproducible B scans. Dempster called his instrument a 'mass spectrometer' but this definition was debated by Aston.[16] In fact, using R from Equation (1.2) one can obtain:

$$R = mv/Bz \tag{1.4}$$

This relationship shows that all the ions entering the magnetic field and having the same charge and the same *momentum*, follow a circular pathway with an equal radius R, independently from their mass, while ions with different *momenta* follow pathways of different radii. For this reason, Aston suggested that the most appropriate term for Dempster's instrument would be 'momentum spectrometer' and not 'mass spectrometer'. However, for ions generated inside the ion source, Equation (1.3) is valid and the term 'mass spectrometer' is appropriate.

The physical application of mass spectrometers for the determination of natural isotope ratios and accurate mass of different nuclides gave rise to the development of instruments with high performance, in particular with increasing resolution. This led to the design of instruments based on the use of magnetic and electrostatic sectors.[17] The researchers engaged in these developments determined that the resolution is mainly affected by four different factors:

(i) ion beam spatial divergence;
(ii) kinetic energy distribution of ions with the same m/z value;
(iii) the curvature radius of the ion pathway inside the magnetic field;
(iv) the width of the ion source and collector slit.

The ion beam emerging from an ion source is, in general, inhomogeneous either in direction or in kinetic energy. It means that the ion beam is partially divergent and consequently it enters in the analyser region with directions inside a θ angle. With respect to kinetic energy distribution, it must be taken into account that not all ions generated inside the ion source experiment the same accelerating field, the potential being inhomogeneous inside the source itself and, considering that $mv^2/2 = zV$, the V inhomogeneity is reflected in the kinetic energy inhomogeneity.

Both these negative aspects are corrected by the use of magnetic and electrostatic sectors.

In 1933 Stephens[18] demonstrated that a magnetic sector leads to direction focusing of the ion beam. Just from the descriptive point of view let us consider the trajectory of an ion beam generated by the source S and focused by the magnetic field at the point C (Figure 1.12). The beam pathway is perpendicular to the field and follows the curve of radius *r*. If an ion enters the magnetic sector with an angle lower than 90°, it undergoes the field action for a longer time and consequently its deviation will be wider, leading to its focusing at C. If the angle is greater than 90° the residence time inside the magnetic field will be lower: its deviation will be smaller and the ion will again be focused at C. From the qualitative point of view, we can say that a magnetic sector focuses at the same point all ions having the same mass, charge and velocity. Hence, a magnetic sector exhibits not only a separating power, but also a focusing one. For these reasons, instruments employing a magnetic sector as a mass analyser were commonly called 'single focusing instruments', leading to direction focusing of the ion beam.

The use of electrostatic sectors is highly effective to overcome the inhomogeneity in kinetic energy. In these devices the ions are subjected to the action of a radial electrostatic field E with direction perpendicular to B. In the field E, generated by two parallel electrodes of cylindrical section (Figure 1.13), the ions are subjected to the action of a centripetal force zE. Calling mv^2/R their centrifugal force it will be:

$$zE = mv^2/R \tag{1.5}$$

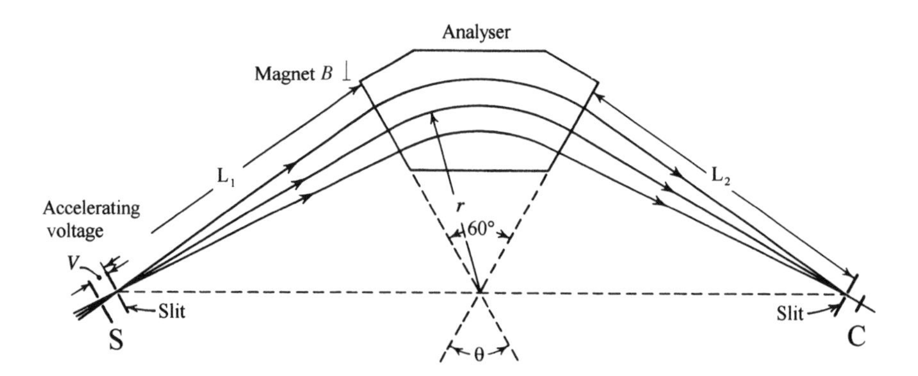

Figure 1.12 Direction focusing action of a magnetic sector

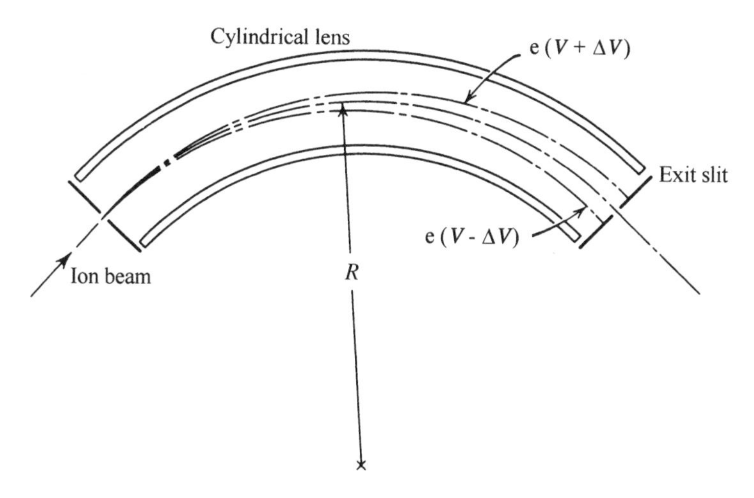

Figure 1.13 Velocity focusing action of an electrostatic sector

which, considering that $zV = \frac{1}{2}mv^2$, leads to the equation:

$$R = 2V/E \tag{1.6}$$

This equation shows that ions accelerated by V and subjected to the action of E follow a circular pathway of radius R, independently from their mass. For E = constant, only the ions with identical kinetic energy pass through the exit slit: the electrostatic sector consequently acts as a kinetic energy filter. Hence analysers employing B and E fields (for example, Figure 1.14) are called 'double focusing instruments'.[19]

Double-focusing instruments exhibit a resolution up to 100 000. Of course the maximum value of resolution corresponds to very narrow slit width and consequently can be achieved in low sensitivity conditions.

1.2.3 Quadrupole Analysers

Quadrupole mass filter[20,21] and quadrupole ion trap[22,23] are currently the mass analysers most widely employed. They were both developed by the Paul group (Nobel Prize for physics in 1989) at Bonn University. The theoretical data of the behaviour of an ion in a quadrupole electrical field is highly complex. Here only a picture of such behaviour is given in order to give the reader a view of what is happening inside these devices.

Both systems start from the same considerations: ions of different m/z values will interact in a different manner with alternate electrical fields

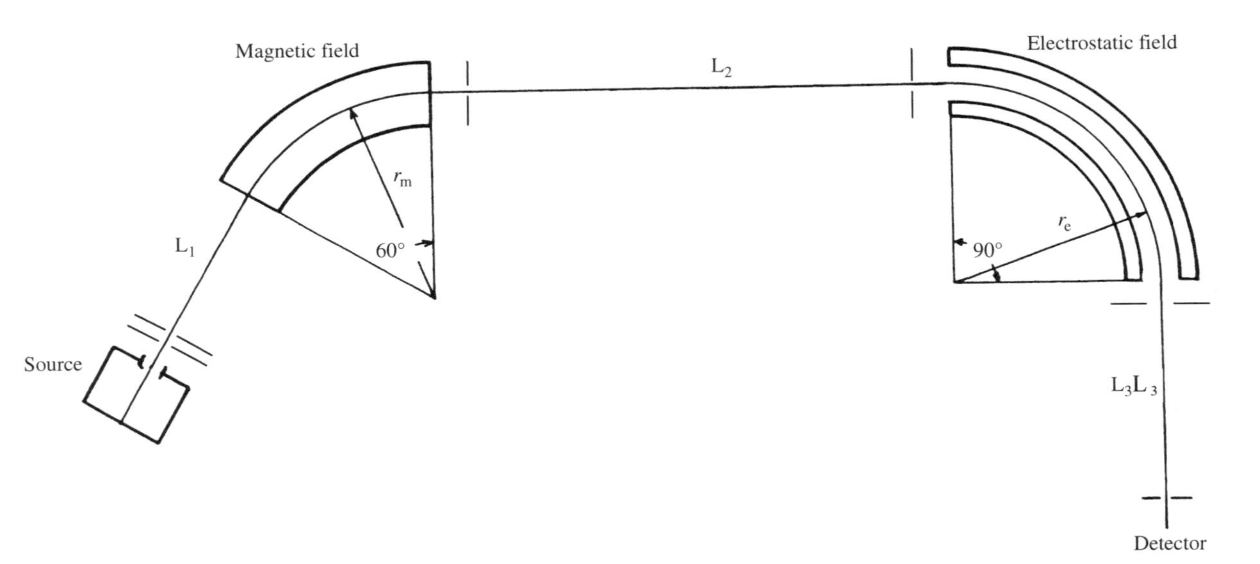

Figure 1.14 Scheme of a double focusing mass spectrometer. L_1, first field-free region; L_2, second field-free region

(radio frequency, RF). Many different devices were developed to study this interaction[21] but, in the analytical world, the quadrupole mass filter and ion trap are the most widely employed.

1.2.3.1 Quadrupole Mass Filter[20,21]

Let us consider an ion ejected from an ion source interacting with a quadrupole field generated by four hyperbolic section rods, as shown in Figure 1.15. On the rods a potential of the type $U \pm V\cos\omega t$ is applied (where U is the direct current potential and $V\cos\omega t$ is the *RF* potential). Ions which enter in this system oscillate in both x and y directions by the action of this field.

Let us consider the parameters that affect this motion; they are the ion m/z value, the U, V, ω values and r_0, i.e. the dimension of the mass filter. If we define now two quantities taking into consideration all these parameters, i.e.:

$$a = 4zU/mr_0^2\omega^2 \tag{1.7}$$

$$q = 2zV/mr_0^2\omega^2 \tag{1.8}$$

it is possible to draw a 'stability diagram' of the device, i.e. to define the a and q values by which the ions follow 'stable' trajectories inside the

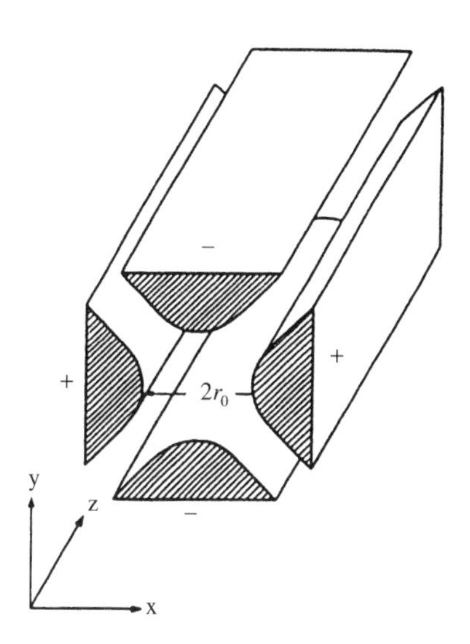

Figure 1.15 Quadrupole mass filter. The ions are injected in the z direction

inter-rod space. Outside this stability diagram the a and q values will be such that the ions will discharge on the rods. The stability diagram usually employed for a quadrupole mass filter is reported in Figure 1.16.

Looking at the a and q definitions it follows that ions with different m/z values exhibit different values of a and q. Furthermore, keeping U, V and ω constant, it is possible to overlap on the same diagram a straight line, whose slope is dependent on the a/q ratio. In fact, the a/q ratio can be calculated as:

$$a/q = (4zU/mr_0^2\omega^2)(mr_0^2\omega^2/2zV) = 2U/V \tag{1.9}$$

which leads to

$$a = (2U/V)q \tag{1.10}$$

This equation represents a straight line in the a, q space, crossing the origin of axes and whose slope is dependent on the U/V ratio. If we choose suitable values, a straight line as that reported in Figure 1.16 can be obtained. It just crosses the apex of the stability diagram and, in these conditions, all the ions exhibiting a and q values out from the apex will follow unstable trajectories. If the V and U values are increased, keeping the U/V ratio constant, ions will increase their a, q values (see Equations 1.7 and 1.8) and, when the straight line portion inside the apex of the stability diagram is reached, they will follow stable trajectories, passing through the rods and reaching the detector placed after the rods themselves. Hence, by scanning both U and V (maintaining the ratio U/V = constant) all the ions can be selectively detected.

The above described behaviour well explains the term 'quadrupole mass filter' of the device. One point to be emphasized is that by varying

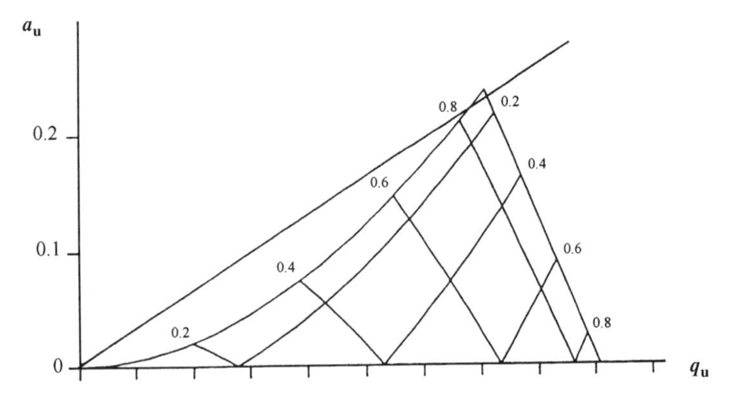

Figure 1.16 q, a stability diagram of a quadrupole mass filter

the *U*/*V* ratio it is possible to vary the straight line slope and consequently its portion inside the stability diagram. Using this approach it is possible to play on the peak width/ion current ratio. In other words, moving the straight line closer to the apex it is possible to obtain a better resolution but the sensitivity may show a significant decrease.

1.2.3.2 Quadrupole Ion Trap[22,23]

The Paul ion trap is constituted by three electrodes arranged in a cylindrical symmetry (Figure 1.17). When a suitable $U + V\cos\omega t$ potential is applied on the intermediate electrode (ring electrode) and the two end-cap electrodes are grounded, a quadrupolar field is generated and the ions inside the trap follow trajectories confined in a well defined space region. Even in this case the behaviour of the ion trap can be described by defining *a* and *q* quantities analogous to those given for the quadrupole mass filter, leading to a stability diagram (Figure 1.18).

Most ion traps commercially available use, as potential applied to the ring electrode, only a RF voltage ($V\cos\omega t$). Consequently the stability

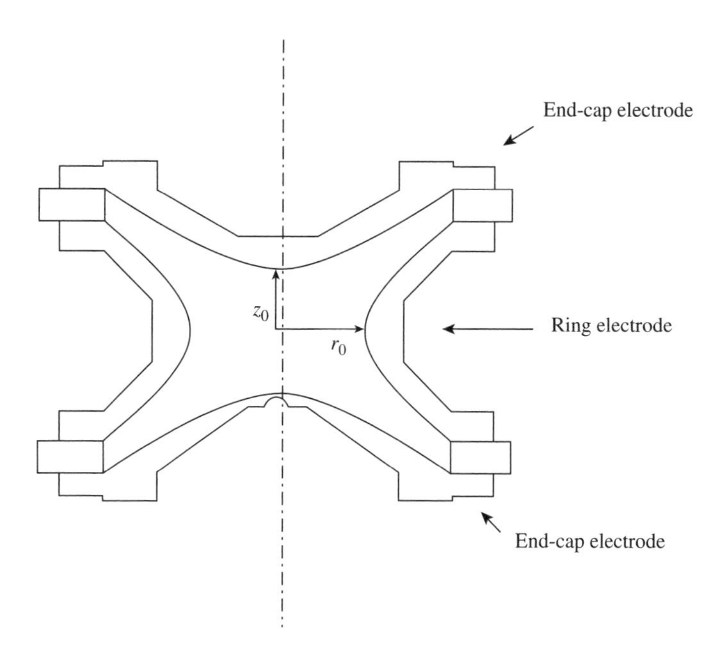

Figure 1.17 Quadrupole ion trap

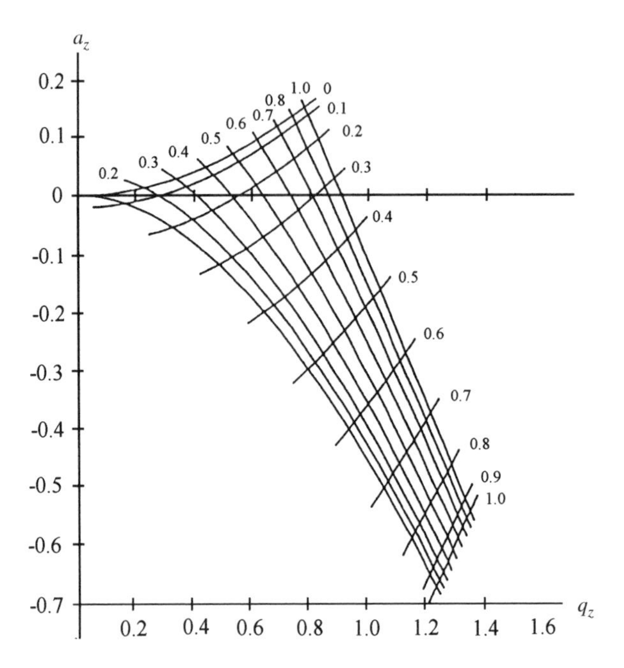

Figure 1.18 q, a stability diagram of a quadrupole ion trap

diagram related to the a, q pairs is in this case just a 'stability line' related to the q values only:

$$q = -4zV/mr_0^2\omega^2 \tag{1.11}$$

From this equation it follows that, for constant values of V, r_0 and ω, ions of different m/z values exhibit different q values and one can imagine different ions lying on the q axis at q values inversely proportional to their m/z values (Figure 1.19). If m/z, V, r_o and ω values are appropriate, all the ions remain trapped inside the device. By scanning

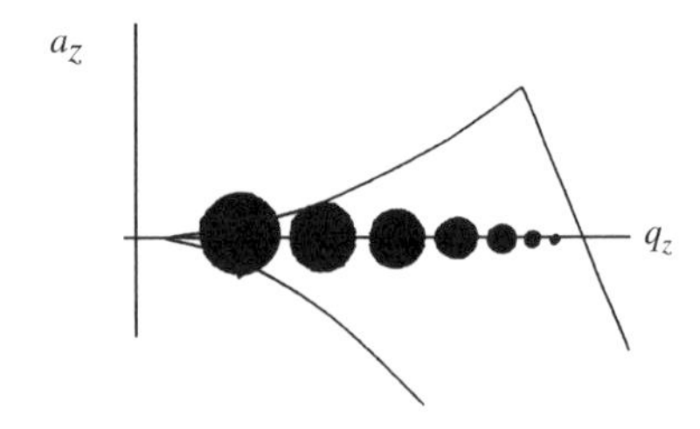

Figure 1.19 Portion of the stability diagram of a quadrupole ion trap. In the absence of U (DC) component, decreasing q values correspond to ions of increasing m/z values

V, the q values of the various ions increase and once their q value reaches the limit of the stability diagram their trajectory increases along the z axis: they are ejected from the trap and detected.

It is to be emphasized that the quadrupole ion trap has specific behaviour that makes it unique. First of all MS/MS experiments (which will be discussed in detail in Section 1.5.3) are easy to perform and the fragmentation yield is so high that sequential collisional experiments (MS^n) can be performed.[23] Secondly, the resolution available by ion trap has been proven to be up to 10^6.[23] Unfortunately, instruments with these latter performances are not still available and only resolution of a few thousand are present in commercial instruments, employed only for partial portions of the mass spectrum. Furthermore two other points are worth noting: on the one hand, the theoretical high mass range available, obtained by the use of a supplementary RF voltage (ion trap with mass range up to m/z 8000 are available); on the other, the high sensitivity of the device (ions produced are not lost during the scanning).[23] These facilities have led to the production of an ion trap with powerful performances for gas chromatography/mass spectrometry (GC/MS) and LC/ESI(APCI)/MS systems.

1.2.4 Time-of-flight

Time-of-flight (TOF)[24] is surely, from the theoretical point of view, the simplest mass analyser. In its 'linear' configuration it consists only of an ion source and a detector, between which a region under vacuum, without any field, is present. Ions with a m/z accelerated by the action of a field V acquire a speed v. The potential energy will be equal to the acquired kinetic energy (Equation 1.1). If, from this equation, we put in the v value:

$$v = [(2zV/m)]^{1/2} \tag{1.12}$$

it is easy to observe that ions of different m/z values exhibit different speeds, inversely proportional to the square root of their m/z values. If the ion follows a linear pathway inside a field-free region of length l, considering that $v = l/t$, i.e. $t = l/v$, it follows that:

$$t = l(m/2zV)^{1/2} \tag{1.13}$$

This equation shows that ions of different m/z value reach the detector at different times, proportional to the square root of their m/z value. For this reason this device is called 'time-of-flight' (Figure 1.20).

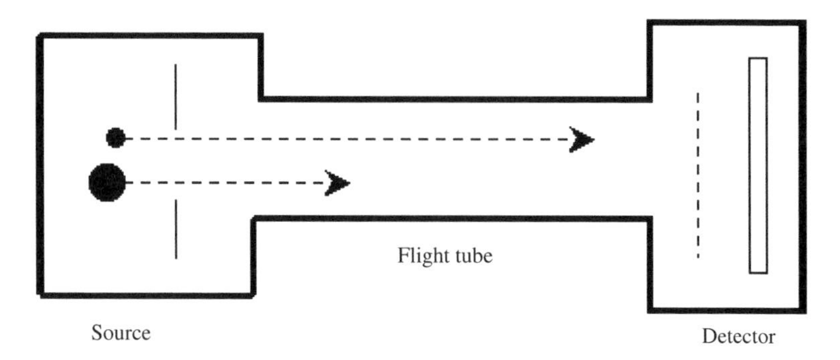

Figure 1.20 Scheme of a linear TOF mass analyser

The calibration of the time scale with respect to *m/z* value can be easily obtained by injection of samples of known mass.

Of course, this analyser cannot operate in a continuous mode as the sector and quadrupole ones. In this case, an ion pulsing phase is required: the shorter the pulse, the better defined is the mass value and the peak shape.

Furthermore, the ions emerging from the source are usually not homogenous with respect to their speed (this effect mainly arises from the inhomogeneity of the acceleration field). Of course, a distribution of kinetic energy will reflect immediately on the peak shape and wide kinetic energy distribution will lead to enlarged peak shape, with the consequent decrease in resolution. To overcome this, different approaches have been proposed and that usually employed consists of a reflectron device. As shown in Figure 1.21, the reflectron is constituted by a series of ring electrodes and a final plate. The plate is placed at a few hundred volts over the V values employed for ion acceleration. By using a series of resistors, the different ring electrodes are placed at decreasing potentials. When an ion beam with kinetic energy $E_k \pm \Delta E_k$ interacts with this field, the ion with excess kinetic energy $(E_k + \Delta E_k)$

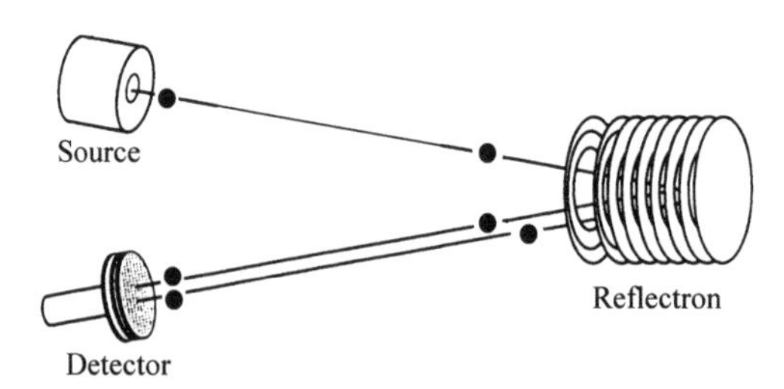

Figure 1.21 Scheme of reflectron TOF

will penetrate that field following a pathway longer than that followed by ions with mean kinetic energy E_k. In contrast, ions with lower kinetic energy will follow a shorter pathway. This phenomenon leads to a thickening of the ion arrival time distribution with a consequent, significant increase in mass resolution.

Nowadays, TOF systems with resolution up to 20 000 are commercially available.

1.3 GC/MS[25]

The coupling of a mass spectrometer with a gas chromatographic system was realized in the 1960s and immediately the scientific community recognized the high power of the system. In the beginning the only chromatographic columns available were packed ones, operating with carrier gas flows in the order of 10 mL/min. Considering the simple relationship between pumping speed (P, L/s) and gas flow (f, mL/min) corresponding to the maintenance of a vacuum in the order of 10^{-5} Torr:

$$P = 10^3 f \tag{1.14}$$

It follows that a direct coupling of a packed column with a mass spectrometer is practically impossible and consequently the use of a He separator was necessary. This led, at the beginning of GC/MS development, to severe limitations of the system, which completely disappeared when capillary columns were introduced. Nowadays, GC/MS systems are robust, reliable, sensitive and highly specific instruments. Their main power is due to the fact that they can be effectively employed for qualitative and quantitative analyses. In most systems, the chromatographic column directly reaches the ion source, usually operating in EI and CI conditions.

The system can operate in three different modes, briefly described below.

1.3.1 Total Ion Current (TIC) Chromatogram

The chromatogram is obtained by sequential MS scanning (typical times are in the order 0.2–0.5 s); the data system takes the sum of the signal due to different ions present in each spectrum (TIC) and plots the signal so obtained with respect to the analysis time. The chromatograms are consequently related to the TIC due to the background (baseline) and to the different components eluting from the chromatographic column (chromatographic peaks).

For quantitative analysis it must be taken into consideration that different compounds exhibit a different ionization yield and consequently lead to a different TIC signal. For such a reason, the use of an internal standard (IS) is essential to obtain reliable quantitative data. These aspects will be discussed in detail in Chapters 2 and 3.

For the qualitative point of view, once the data file corresponding to the TIC chromatogram is obtained, it is easy to obtain the mass spectra of various components by just looking at the data with respect to their retention time.

1.3.2 Reconstructed Ion Chromatogram (RIC)

The data file obtained by the TIC approach can be validly employed to confirm the presence of and to quantify species of interest present in a complex mixture. Using the data it is possible to select one or more ions of interest and reconstruct a chromatogram related to their *m/z* value only. By this approach a high specificity can be achieved and practically an increase of chromatographic resolution can be obtained.

For example, in Figure 1.22 the TIC chromatogram shows a wide peak in the range 12.5–17 min, but when the ion chromatograms

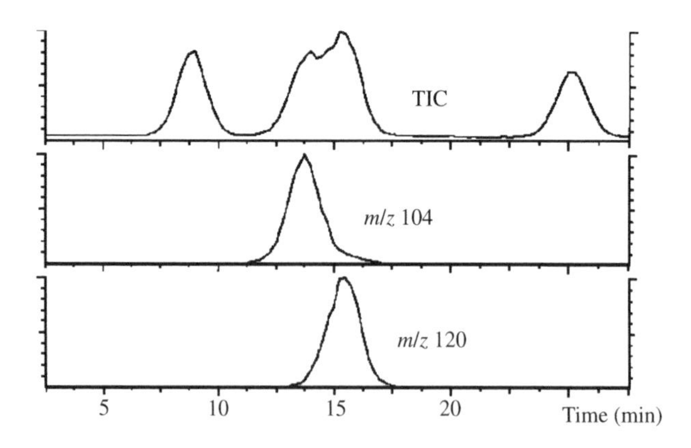

Figure 1.22 Gas chromatograms obtained by TIC and ion current related to ions at *m/z* 104 and 120

relating to *m/z* 104 and 120 values are reconstructed, it is found that the peak is due to the partial overlapping of two different components.

1.3.3 Multiple Ion Detection (MID)

In the case of trace analysis, it is useful not to perform the scan of all the spectra, but to consider the different *m/z* values characteristic of the mass spectrum of the compound of interest. Using this approach, a higher speed monitoring of the chromatographic eluate can be obtained and valid quantitative data are obtained even for species present at ppb level. This approach can be applied either on mass units or on accurate mass values, leading to an increase of specificity of the method.

1.4 LC/MS[26]

In the beginning, the coupling of LC with traditional (EI, CI) sources was unsuccessful, due to the high solvent level present in LC eluate. Some attempts were made to remove the solvent by 'moving belt' systems and to use the solvent itself as reactant for low pressure CI experiments. The situation changed with the development of atmospheric pressure ion sources, such as in APCI, ESI and, to a minor extent, APPI.

Nowadays these systems are widely employed and allow the application of mass spectrometry in fields (especially biological and biomedical) which only a few decades were completely off limits.

A typical arrangement for LC/MS measurements is shown in Figure 1.23. It consists of a LC pump, LC column, APCI or ESI source and mass spectrometer. The LC pump system and the LC column must be chosen considering the source employed. In fact, due to the different ionization mechanisms, each source has its own optimum eluent flow rate and solvent polarity. The APCI source operates properly with eluent flow rates higher than those employed for ESI and is compatible with a

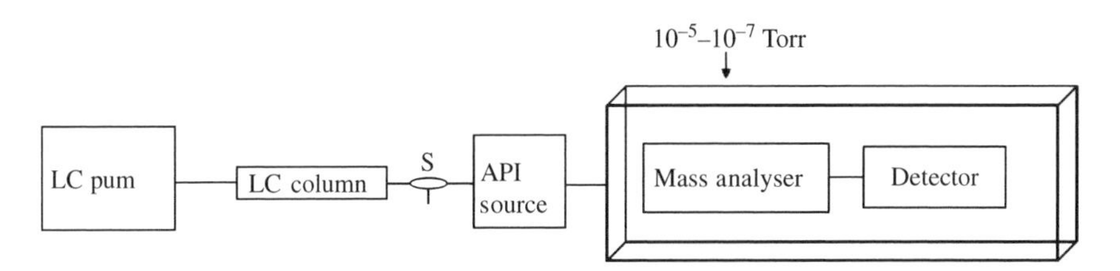

Figure 1.23 Scheme of the LC/MS coupling

nonpolar mobile phase. Typical eluent flow rates employed with an APCI source are in the range 4.0–0.5 mL/min: below 0.4 mL/min an unstable analyte signal can be observed due to nonreproducible discharge processes. This implies the use of a conventional LC pump and normal (3–4.6 mm ID) and narrow-bore (1–2 mm ID) LC column.

A modern ESI source can work with eluent flow rate up to 1 mL/min, even if the optimum value is about 0.2–0.3 mL/min. A normal LC column can be used by splitting the eluent before the entrance in the ESI source, while narrow-bore LC columns are normally employed without splitting. The most recent ESI developments lead to micro-ESI and nano-ESI, employed for high sensitivity measurements and applications where sample amounts are limited (i.e. proteomic, pharmacokinetic studies, etc.). Flow rates in the range 100–10 000 nL/min are normally used and can be obtained by a capillary LC pump, equipped with a capillary (0.15–0.8 mm ID) and nano (20–100 μm ID) LC column.

As in the case of GC/MS, TIC chromatogram, RIC and MID can be obtained.

1.5 MS/MS[27]

MS/MS is based on the use of a first mass analyser, employed to select ions of interest, and a second mass analyser devoted to the analysis of the decomposition products of the preselected ions.

The MS/MS experiment can be subdivided into four different steps:

(i) ion generation;
(ii) ion selection;
(iii) selected ion decomposition;
(iv) mass analysis of the selected ion decomposition products.

These steps can occur in different space regions and in this case the experiment is called MS/MS ‘in space’, or in the same space region and consequently they must be time separated. The latter approach is called MS/MS ‘in time’.

1.5.1 MS/MS by Double Focusing Instruments

The first experiments of MS/MS were generated by the study of the naturally occurring decomposition of selected ions in the region between the magnetic and electrostatic sector of a double focusing instrument (see, for example, Figure 1.14).[28] The ions of interest, produced in the ion source, were selected by fixing the related *B* value. The decomposition products were analysed by scanning the electrostatic sector. In order to increase the yield of ion fragmentation a 'collision cell', i.e. a small box in which a collision gas (typical pressures in the range 10^{-2}–10^{-3} Torr) was inserted in that region, and a fantastic increase of product ion abundance number and abundance was observed. The method was called collisional-induced decomposition (CID) and in the first years of its application its main use was in the field of structural analysis of gaseous ions.

However, in the 1980s some papers appeared, showing the power of the method in the analysis of compounds of interest present in complex mixtures. In fact, without the use of any separative method, it is possible, by direct introduction of a complex mixture and selection of the ion characterizing the analyte, to determine its presence and, in some cases, to obtain some quantitative data.

1.5.2 MS/MS by Triple Quadrupoles

The real introduction of the MS/MS system in the analytical world started with the development of triple quadrupole (QQQ) systems,[29] shown schematically in Figure 1.24. The ions of interest (M^+), produced by the suitable ionization method, are selected by Q_1, by choosing the appropriate *U* and *V* values. The collision gas is injected in Q_2, which operates in RF only (i.e. it behaves as an ion lens). The ions originating by collisionally induced decomposition of M^+ are analysed by Q_3.

This instrumental arrangement allows a wide series of collisional experiments to be performed, among which the most analytically relevant are:

(i) product ion scan: identification of the decomposition product of a selected ionic species;

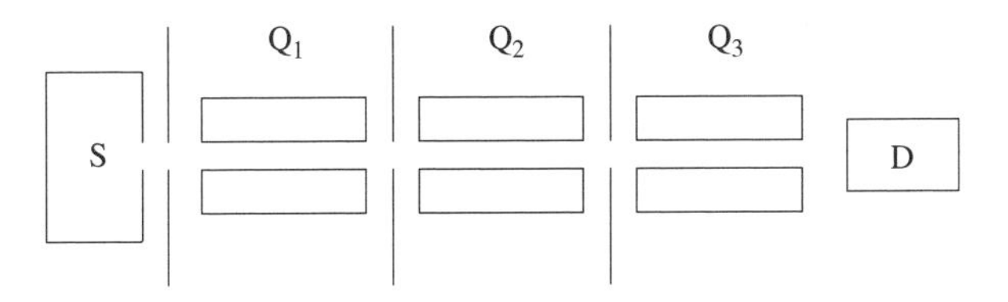

Figure 1.24 Scheme of a triple quadrupole (QQQ) system for MS/MS experiments

(ii) parent ion scan: identification of all the ionic species which produce the same fragment ion;
(iii) neutral loss scan: identification of all the ionic species which decompose through the loss of the same neutral fragment.

The collisional phenomena occurring in a triple quadrupole (as well as in sector machines) lead to the production of an ion population with a wide internal energy distribution, due to the statistics of the preselected ion – target gas interactions. Hence, various decomposition channels, exhibiting different critical energies, can be activated and the resulting MS/MS spectrum is, in general, particularly reach of peaks and, consequently, of analytical information.

What are the parameters which one can vary in a MS/MS experiment by QQQ? Two parameters are the nature of the target gas (the larger the target dimension, the higher the internal energy deposition on the preselected ion: in other words, Ar is more effective than He) and its pressure (the higher the pressure, the higher the probability of multiple collisions leading to increased decompositions: of course the pressure must not exceed the limit compromising the ion transmission!). But, over all, the kinetic energy of colliding ions, which can be varied by suitable electrostatic lenses placed between Q_1 and Q_2, plays a fundamental role in MS/MS experiments.

1.5.3 MS/MS by Ion Traps

More recently, ion trap showed interesting behaviour for MS/MS experiments.[23] It represents an example of MS/MS 'in time'. In fact, the sequence ion isolation – collision – product ions analysis is performed in the same physical space and consequently must be time-separated. A typical sequence is reported in Figure 1.25.

The ions are generated inside the ion trap (or injected in the trap after their outside generation) for a suitable time, chosen in order to optimize the number of trapped ions (a too high ion density leads to degraded data due to space-charge effects). The ions inside the trap exhibit motion frequency depending on their *m/z* values. The ion selection phase is achieved by the application, on the two end-caps, of a supplementary RF voltage with all the ion frequencies but the ion of interest one. In these conditions all the undesidered ions are ejected from the trap and only that of interest remains trapped. The collision of the preselected ion is again performed by resonance with the supplementary RF field with a frequency

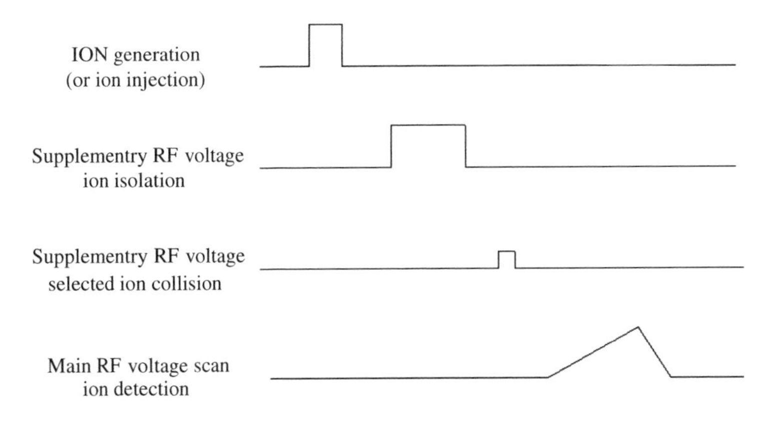

Figure 1.25 Sequential pulses of supplementary and main RF voltages employed to perform MS/MS experiments by ion trap

corresponding to that of ion motion, but with an intensity such to maintain the ion trajectory inside the trap walls. The ion collides with the He atoms, present in the trap as buffer gas and, once sufficient internal energy is acquired, it decomposes: the product ions so generated remain trapped and by the main RF scan they are ejected from the trap and detected.

It should be emphasized that the collisional data obtained by ion trap are quite different from those achieved by QQQ. In fact in this case the energy deposition is a step-by-step phenomenon.[30] Each time that the ion is accelerated by the supplementary RF field up and down inside the trap, it acquires, through collision with He atoms, a small amount of internal energy. When the internal energy necessary to activate the decomposition channel(s) at the lowest critical energy is reached, the ion fragments. In other words, while in the case of QQQ the wide internal energy distribution from collisional experiments leads to the production of a large set of product ions, in the case of ion trap only a few product ions are detected, originating from the decomposition processes at lowest critical energy.

This aspect could be considered negative from the analytical point of view: in fact a better structural characterization can be achieved by the presence of a wider product ions set. But it can be easily and effectively overcome by the ability of ion trap to perform multiple MS/MS experiments. In fact the sequence shown in Figure 1.25 can be repeated by selection, among the collisionally generated product ions, of an ionic species of interest, its collision and the detection of its product ions (MS^3). This process can be repeated more times (MS^n), allowing on the one hand to draw a detailed decomposition pattern related to low energy

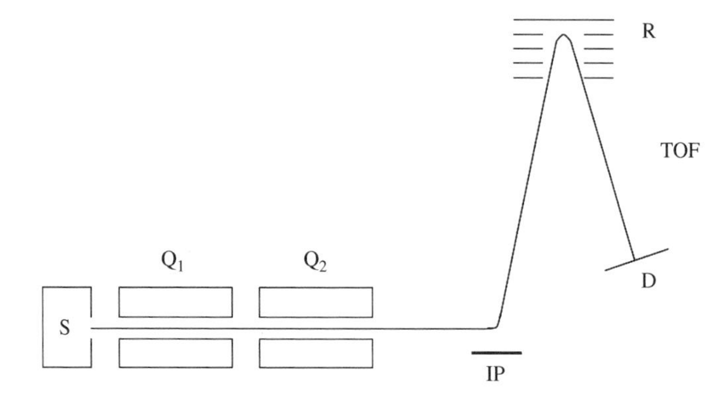

Figure 1.26 Scheme of a Q-TOF system. S, ion source; Q_1, quadrupole for the selection of ionic species of interest; Q_2, collision region (Q_2 operates in RF only); IP, ion pusher; R, reflectron; D, detector

decomposition channels, and on the other to obtain fragment ions of high diagnostic value from the structural point of view (and hence analytically highly relevant). Hence, by ion trap it is possible to perform MS^n experiments, which cannot be obtained by the QQQ approach.

1.5.4 MS/MS by Q-TOF

Recently a new MS/MS instrument has become available, exhibiting a specificity higher than that achieved by QQQ or ion trap systems. It is based on a 'hybrid' configuration, employing mass analysers based on different separation principles. The system, usually called Q-TOF is shown schematically in Figure 1.26. The ion of interest, generated in the source S, is selected by the quadrupole mass filter Q_1. Collisions take place in Q_2 (a quadrupole operating in RF only). The product ions are analysed by a TOF analyser. Considering the high resolution conditions available by TOF, by this approach the accurate masses of the collisionally generated product ions (as well as of the precursor) can be easily obtained, allowing the determination of their elemental composition. This possibility leads to a significant increase in the specificity of MS/MS data.

REFERENCES

1. T. D. Mark and G. H. Dunn (Eds) *Electron Impact Ionization*, Springer–Verlag, Wien (1985).

2. A. G. Harrison, *Chemical Ionization Mass Spectrometry*, CRC Press, Boca Raton (1983).
3. S. G. Lias, J. E. Bartness, J. F. Liebamnn, J. L. Holmes, R. D. Levin and W. G. Mallard, *J. Phys. Chem. Ref. Data*, **17** (Suppl. 1) (1988).
4. H. D. Beckey, *Principles of Field Ionization and Field Desorption Mass Spectrometry*, Pergamon, London (1975).
5. M. Barber, R. S. Bordoli, G. J. Elliott, R. D. Sedgwick and A. N. Tyler, *Anal. Chem.*, **54**, 645–657A (1982).
6. A. P. Bruins, *Mass Spectrom. Rev.*, **10**, 53–77 (1991).
7. (a) M. Yamashita and J. B. Fenn, *J. Phys. Chem.*, **88**, 4451–4459 (1988); (b) M. Yamashita and J. B. Fenn, *J. Phys. Chem.*, **88**, 4671–4675 (1988).
8. D. B. Robb, T. R. Covey and A. P. Bruins, *Anal. Chem.*, **72**, 3653–3659 (2000).
9. M. Karas, D. Bahar and U. Griessmann, *Mass Spectrom. Rev.*, **10**, 335–357 (1991).
10. R. J. Pfeiffer and C. D. Hendricks, *AIAA J.*, **6**, 496–502 (1968).
11. G. L. Taylor, *Proc. R. Soc. London Ser. A*, **280**, 383–387 (1964).
12. M. Dole, L. L. Mack, R. L. Hines, R. C. Mobley, L. D. Ferguson and M. B. Alice, *J. Chem. Phys.*, **49**, 2240–2249 (1968).
13. J. V. Iribarne and B. A. Thomson, *J. Chem. Phys.*, 2287–2294 (1976).
14. P. Traldi and E. Marotta in *Advances in Mass Spectrometry*, Vol. 16, A. E. Ashcroft, G. Brenton and J. J. Monagan (Eds), Elsevier, Amsterdam (2004), pp. 275–293.
15. A. J. Dempster, *J. Phys. Rev.*, **11**, 316–320 (1918).
16. F. W. Aston, *Nature*, **127**, 813–820 (1931).
17. (a) J. H. Beynon in *Mass Spectrometry and its Applications to Organic Chemistry*, Elsevier, Amsterdam (1960), pp. 4–27; (b) F. A. White and G. Wood in *Mass Spectrometry Applications in Science and Engineering*, John Wiley & Sons, Inc., New York (1986), pp. 51–66.
18. W. E. Stephens, *Phys. Rev.*, **45**, 513–518 (1934).
19. (a) F. A. White, F. M. Rourke and J. M. Sheffield, *Appl. Spectr.*, **12**, 46–52 (1958); (b) T. Wachs, P. F. Bente and F. W. McLafferty, *Int. J. Mass Spectrom. Ion Phys.*, **9**, 333–341 (1972); (c) J. H. Beynon, R. G. Cooks and J. W. Amy, *Anal. Chem.*, **45**, 1023 (1973); (d) H. Hintenberger and L. A. König, *Z. Naturforsch. A*, **12**, 443–452 (1957); (e) R. P. Morgan, J. H. Beynon, R. H. Bateman and B. N. Green, *Int. J. Mass Spectrom. Ion Phys.*, **28**, 171–191 (1978).
20. (a) W. Paul and H. Steinwedel, Ger. Pat. 944, 900 (1956); US Pat. 2, 939, 952 (1960); (b) W. Paul and H. Steinwedel, *Z. Naturforsch. A*, **8**, 44 (1953); (c) W. Paul, H. P. Reinhard and U. von Zahn, *Z. Phys.*, **152**, 153 (1958).
21. P. H. Dawson (Ed.) *Quadrupole Mass Spectrometry and its Applications*, Elsevier, Amsterdam (1976).
22. R. E. March and R. J. Hughes, *Quadrupole Storage Mass Spectrometry*, John Wiley & Sons, Inc., New York (1989).
23. R. E. March and J. F. J. Todd (Eds) *Practical Aspects of Ion Trap Mass Spectrometry*, Vols I–III, CRC Press, Boca Raton (1995).
24. (a) W. C. Wiley and I. H. McLaren, *Rev. Sci. Instr.*, **26**, 1150–1157 (1955); (b) R. J. Cotter, *Anal. Chem.*, **64**, 1027A–1039A (1992); (c) B. A. Mamyrin, V.I. Karataev, D. V. Shmikk and V. A. Zagulin, *Sov. Phys.– JETP*, **37**, 45–48 (1973); (d) B. A. Mamyrin, *Int. J. Mass Spectrom. Ion Proc.*, **131**, 1–19 (1994).
25. J. Abian and E. Gelpi in *Mass Spectrometry in Biomolecular Sciences*, R. Caprioli, A. Malorni and G. Sindona (Eds), NATO ASI Series, Series C: Mathematical and

Physical Sciences, Vol. 475, Kluwer Academic Publisher, Dordecht (1996), pp. 437–460.

26. (a) B. E. Erickson, *Anal. Chem.*, **72**, 711A–716A (2000); (b) W. M. Niessen, *J. Chromatogr. A*, **856**, 179–197 (1999).
27. F. W. McLaffery (Ed.) *Tandem Mass Spectrometry*, John Wiley & Sons, Inc., New York, (1983).
28. R. G. Cooks, J. H. Beynon, R. M. Caprioli and G. R. Lester, *Metastable Ions*, Elseveier, Amsterdam (1973).
29. R. A. Yost and C. G. Enke in *Tandem Mass Spectrometry*, F. W. McLaffery (Ed.), John Wiley & Sons, Inc., New York (1983), pp. 175–195.
30. J. Gronova, C. Paradisi, P. Traldi and U. Vettori, *Rapid Commun. Mass Spectrom.*, **4**, 306–313 (1990).

2

How to Design a Quantitative Analysis

Chemical analyses are measure procedures and consequently the general concepts of metrology can be applied to them. According to metrology, in order to obtain reliable results from a measurement, it must refer to certified standards. This allows a valid comparison of measurements performed in different conditions and at different times. Furthermore, two different measurements are compatible only when the related uncertainty value is provided.

In the past most attention was focused on the reproducibility of measurements but nowadays the interest is mainly focused on the possibility of comparing the results obtained by different laboratories. This interest originates from various aspects of globalization, the increase and liberalization of work trade, the achievement of uniformity of analysis and medical treatment and the growing, common interest in environmental problems and planet health.

In this frame, retention of the calibration procedure is essential. It can be defined as 'all the operations which allow to establish, under given conditions, the relationship between the value of a quantity indicated by an instrument or any other measurement system and the corresponding value present in a standard sample'. The results of a calibration procedure are given sometimes as a calibration constant or a series of calibration constants expressed in a calibration diagram. For an instrumental measurement this diagram represents the response of the instrument itself to different values of the quantity under measurement. The curve linking the points so obtained is called the calibration curve. An

Quantitative Applications of Mass Spectrometry I. Lavagnini, F. Magno, R. Seraglia and P. Traldi

uncertainty band should be added to the calibration curve. All these points will be treated in detail in Chapter 4.

The calibration of an instrument is a very specific operation that must take place at a very precise moment. After calibration the instrument can be used; calibration remains constant for some time. Calibration must be checked and sometimes adjustment of the instrument or a new calibration operation become necessary.

As far as chemistry is concerned, the most required measurements are the different forms of concentration (mass concentration, volume concentration, molar fraction and so on). Therefore standard different substances at different concentrations are needed. In general the ideal standard is a sample of a pure substance diluted in known proportions into a proper matrix.

Very often the analytical method cannot guarantee either a complete separation or a complete recovery of the substances present in the material to be analysed, or the absence of interferences. For that reason it is necessary to use a reference matrix as similar as possible to the composition of the material to be analysed.

In the case of particularly complex matrices (both organic and inorganic) reference material is not prepared mixing its own components, but starting from a material similar to the one that has to be analysed (such as sludges or animal tissue, for instance). The determination of the concentration of each different component is done analytically with specific precautions (and often with the help of several and different laboratories).

It is important to highlight the effort made during recent years not only by the researchers belonging to different scientific fields to fix standards fitting all branches of science and technology, but also the effort made by people working in different sectors to adopt a standard terminology and procedures for solving the problems of measurement quality which may be used in different scientific branches such as physics, chemistry, clinical chemistry and laboratory medicine.

After these general considerations on the metrology approach, we will focus our attention on the procedures generally adopted in the project and design of a quantitative analysis.

2.1 GENERAL STRATEGY

The general strategy employed in the design of a quantitative analysis is shown schematically in Figure 2.1. It consists of many different steps,

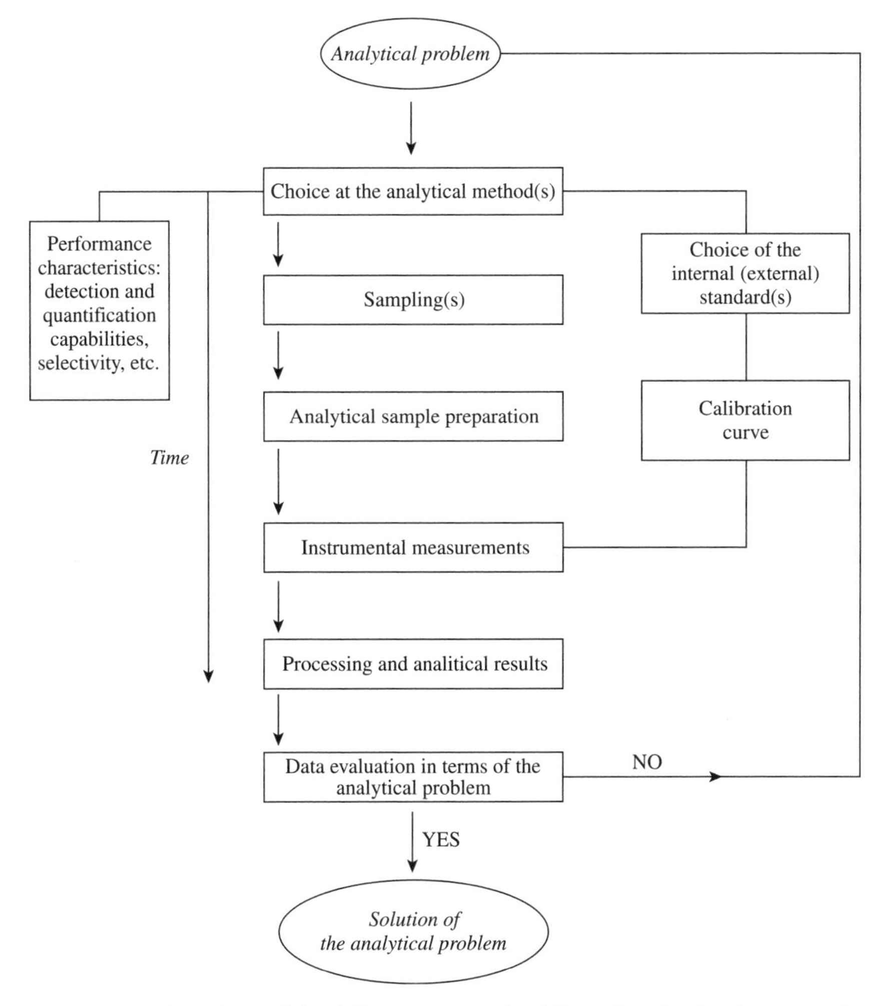

Figure 2.1 Flow chart of the different steps to be followed in the development of a quantitative analysis

each of which is relevant to the achievement of valid results. If, after the development of the procedure, the data obtained are not satisfactory for the solution of the analytical problem, the whole procedure must restart with the use of a different analytical method. On the left side of the flow chart one important parameter, the analysis time, is reported; it becomes particularly important when the number of samples to be analysed is high: in this case a high throughput becomes essential.

In general the cost/benefits ratio of the possible different analytical approaches must be evaluated: the low ratio value obtained for mass spectrometric approaches is not related to the low costs, but rather to the high value of benefits in terms of sensitivity and specificity, as well as to their high throughput.

The analytical methods mainly employed can be subdivided into two broad classes, the chemical-physical and immunometric ones. Mass spectrometry, with all its possible operative configurations, belongs to the former group.

At this point it may be useful to define a series of quantities which give the performance characteristics of an analytical procedure:

Applicability
The applicability pertains to the analyte identity, its concentration range, and the acceptable uncertainty.

Specificity
The specificity is the capability of an analytical method to reveal only one analyte in the presence of other components.

Selectivity
The selectivity is the capability of an analytical method in quantifying analytes in the presence of interfering species.

Precision
The precision is the accordance among independent results obtained in the same operating conditions.

Accuracy
The accuracy evaluates the proximity between the true (or the accepted reference value) and the found value; the inaccuracy comprises the random error and the bias.

Trueness
The trueness is the proximity between the true (or the accepted reference value) and the mean value of several repeated measurements.

Recovery
The recovery is the estimation, in per cent, of the amount of the species recovered in the analytical procedure.

Range
The range is defined as the difference between the minimum and maximum response level.

Linear dynamic range
This is the concentration interval where the signal changes linearly with the analytical concentration.

Critical level
This is the minimum signal which can be distinguished from the background with a defined confidence level.

Limit of detection
This is the lowest quantity or concentration of the analyte which can be detectable with a defined confidence level.

Limit of quantification
This is the lowest quantity or concentration of the analyte whose response is measured with a defined precision.

Sensitivity
This is the slope of the calibration curve.

Resistance
This quantity expresses the capability of the method to give the same analytical result for the same sample even in presence of changes of experimental conditions (e.g. different laboratories, different instrumentation, different days of analysis).

Robustness (or Ruggedness)
This quantity expresses the capability of the method to be not strongly dependent on small changes in operating conditions (e.g. temperature, pH).

2.1.1 Project

The design of a project for a quantitative analysis is generated by an input related to the kind of analyte(s) of interest, its possible quantity and the substrate in which the analyte(s) is present. These three points must be carefully analysed to make the best choices in terms of analyte recovery and instrumental set-up.

2.1.2 Sampling

The sampling must be done in a manner which warrants no change in the analyte composition from the qualitative and quantitative points of views, and the achievement of statistically significant results.

Two sampling approaches can be employed, strongly dependent on the substrate in which the analyte is present. The first is based on on-line, in-field measurements with the instrument present on the site of interest and the sampling is carried out by a device directly connected to the instrument. This is the case, for example, of quadrupole gas analysers: the gaseous substrate (usually air at atmospheric or reduced pressure) is directly injected in an EI source and the gaseous analyte partial pressures are immediately determined by mass analysis. The second approach, the most common one, is based on off-line measurements: in this case the substrate containing the analyte(s) of interest is brought to the laboratory, where analyte extraction procedures and instrumental measurements are performed. A mixed procedure can also be considered, based on the analyte extraction in the field and its analysis in the laboratory. This is the case in which sampling cartridges or solid phase microextraction (SPME) devices are employed.

There are some general rules to be followed for a correct sampling:

(i) the sampling must be carried out by trained and qualified personnel;
(ii) each lot to be analysed must be sampled separately;
(iii) the sampling devices must be appropriate and must guarantee the absence of any contamination;
(iv) the incremental samples must be drawn in different points of the lot or sub-lot;
(v) the aggregate sample is obtained by mixing the incremental samples;
(vi) the laboratory samples are obtained from the homogenized aggregate sample;
(vii) the samples must be suitably stored.

A correct labelling of the laboratory sample is fundamental: it would contain sample name, sampling date and place, and lot and any information useful for the analyst.

2.1.3 Sample Treatment

The analytical sample preparation is a particularly critical point. It consists usually in several steps: analyte extraction from matrix, extract purification and, when necessary, extract concentration, and chemical transformation (derivatization) of the analyte in the form more suitable for the analysis.

An effective development of the sample preparation for the expected analyte levels requires specific knowledge of the chemical and physical properties of the analyte and of the matrix. The different sample extraction procedures commonly employed are summarized in Figure 2.2. Their description is beyond the aim of the present work and the reader can find details on the different approaches in scientific and technical literature.

Samples \ Matrix	Solid	Liquid
Volatile	Gas phase extraction • Static head space (SHS) • Dynamic head space (DHS) Solid phase microextraction (SPME) Supercritical fluid extraction (SFE)	Gas phase extraction • Static head space (SHS) • Dynamic head space (DHS) • Purge & trap (P&T) Solid phase microextraction (SPME) Supercritical fluid extraction (SFE)
None volatile	Solid–liquid extraction • Soxhlet • Ultrasonic extraction • Matrix solid phase dispersion • Accelerated solvent extraction • Microwave assisted extraction Supercritical fluid extraction (SFE)	Liquid–liquid extraction • Separatory funnel • Continuous extraction • Simultaneous distillation extraction (SDE) Solid phase extraction (SPE) Solid phase microextraction (SPME) Stir bar sorptive extraction (SBSE)

Figure 2.2 Different sample extraction procedures commonly employed

2.1.4 Instrumental Analysis

A quantitative analysis based on MS methods is usually carried out by hyphenated techniques, mainly GC/MS and LC/MS. There are some similarities and some differences among the problems arising from the two approaches and some of these aspects will be discussed in detail in Chapter 3. At this stage the same treatment can be used for both approaches. In fact, the same methodologies are employed to obtain the related chromatogram and the quantitative measurements are always related to the relationship between instrumental response and sample quantity (or concentration).

2.1.4.1 Quantitative Analysis by GC/MS and LC/MS

GC/MS and LC/MS measurements can be performed in different operating conditions:

(i) Full scan. Full scan spectra (in the mass range of interest) are continuously recorded, with typical scanning speed in the order 0.2–1 s. The chromatographic signal is obtained by the sum of the MS signals due to all the ions detected in the selected mass range or to selected ions (full scan – selected extraction ion profile) when resolution in the *m/z* domain is utilized to improve inadequate resolution in the time domain. It should be emphasized that molecules of different structures exhibit different ionization yields and consequently the number of ions (and the chromatographic peak areas) can be different for different compounds present in equal molar ratio. For this reason the use of a calibration procedure becomes essential.

(ii) Selected ion monitoring (SIM) or multiple ion detection (MID). The mass spectrometer operates by jumping on *m/z* values highly diagnostic for the analyte of interest. By this approach a drastic reduction of the chemical background is obtained; in other words, all the components for which the diagnostic ions are not present in their mass spectra and possibly interfering with the analyte [with identical or similar retention time (r.t.) values] are not detected. The SIM (or MID) analysis can be performed by operating on integer *m/z* values (i.e. in low resolution conditions) or, as will be described in Chapter 3, on accurate mass values (i.e. in high resolution conditions). Considering that isobaric ions of different elemental composition lead to different values of accurate mass, it follows that by the latter approach a clear increase in specificity and selectivity is obtained, in their turns reflecting on an increase in detection limits and sensitivity. However, also in this case, the use of an internal standard is essential.

As an example, the partial TIC chromatogram of a urine sample extract is shown in Figure 2.3(a). The peak at r.t. ~19 min corresponds to bis-tetramethylsilane derivatives of testosterone, whose EI mass spectrum is characterized by the presence of high abundant ions at *m/z* 432 ($M^{+\cdot}$), 417 ($[M - CH_3]^+$) and 301 ($[M - C_6H_{15}OSi]^+$). The absolute quantity of testosterone was determined in the order of 100 pg. The selection and monitoring of the ions at *m/z* 432, 417 and 301 led to a chromatogram with only the peak due to testosterone. As shown in Figure 2.3(b), an increase in detection limits and sensitivity is obtained. The selected ion signal-to-noise ratio for a sample containing 5 pg is analogous to that shown in Figure 2.3(a) (100 pg).

(iii) MS/MS and MS^n. A further increase in specificity and selectivity can be obtained by collisional experiments on ionic species diagnostic for the analyte and the recording of collisionally produced fragment ions. It is easy to achieve the selection of a diagnostic ion, and the production of a diagnostic fragment practically eliminates all the interfering species. This method

(a)

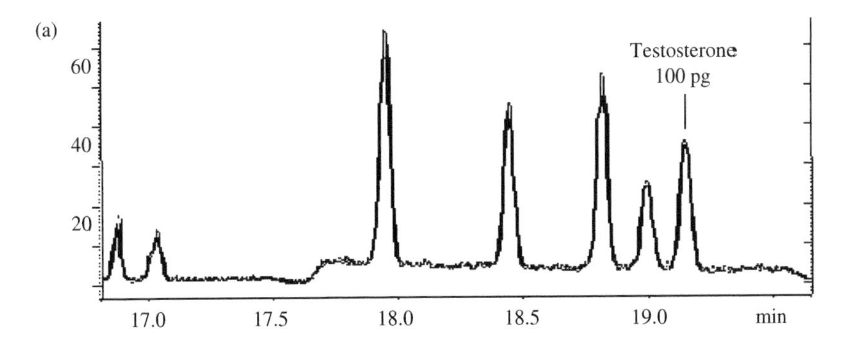

(b)

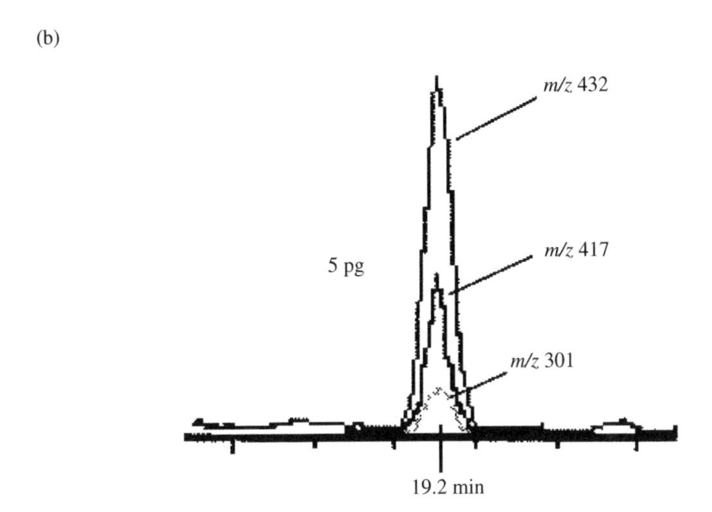

Figure 2.3 (a) TIC gas chromatogram of a urine sample containing about 100 pg/μL of testosterone. (b) MID chromatogram of a urine sample containing 5 pg/μL **of testosterone**

can be further improved by the use of sequential collision experiments (MS^n), such as those possible with ion trap instruments. Also, for this approach the use of an internal standard is essential. The recording of the ion current related to collisionally generated ions is usually called selected reaction monitoring (SRM).

2.1.4.2 Calibration Curves

A most rigorous description of this topic is given in Chapter 4. At this point, it is useful to give a description on how to obtain a valid calibration curve that can be successfully used for quantitative measurements.

Quantitative measurements are usually based on the construction of a relationship between the instrumental response and the analyte concentration/amount. The slope of this curve is called the sensitivity or calibration factor S. In the case of GC/MS and LC/MS measurements the instrumental response is the analyte peak area (as obtained by either TIC or selected ion current). This relationship would be, in some concentration ranges, linear but, due to experimental and/or observational errors, the experimental points are (hopefully weakly) scattered about the straight line.

Apart from random fluctuations, the relationship between the instrumental response and the concentration/amount may be disturbed by interferences, matrix effects and by the simultaneous presence of these two effects. We here disregard the interference problem as we trust that the use of a hyphenated instrument which combines time and *m/z* resolution can eliminate it. The matrix effect, which modifies the sensitivity of the analysis, must be still considered and can be handled by the addition of standards of the analyte to the sample (Standard Addition Method). However, when the matrix effect is negligible or when the matrix can be satisfactorily reproduced, the usual 'external' calibration procedure can be successfully adopted.

The Standard Addition Method can be used to determine the unknown level of a known analyte in a unreproducible matrix. By this approach a plot, as shown in Figure 2.4, is obtained. Once prepared a standard solution of the analyte at known concentration C_{st}, a constant volume of the unknown solution V_u is added to each of four (or more) flasks of equal volume V_f. At this point increasing (and well measured) volumes of the standard solution V_{st} are added to each flask and the solutions are brought to the same volume by adding solvent and then measured. By this approach it is easy to calculate the unknown concentration C_0 of the analyte. Considering that the analyte concentration in a flask can be expressed as:

$$C_u = (C_0 V_u + C_{st} V_{st})/V_f = C_0 V_u/V_f + C_{st} V_{st}/V_f \tag{2.1}$$

and that the response R is the product of the calibration factor S and the concentration, Equation (2.1) can be written as:

$$R = S \cdot C_0 V_u/V_f + S \cdot C \tag{2.2}$$

where $C = C_{st} V_{st}/V_f$.

Equation (2.2) represents a straight line with intercept equal to $S \cdot C_0 V_u/V_f$. The null response is obtained at $C = -C_0 V_u/V_f$ (Figure 2.5)

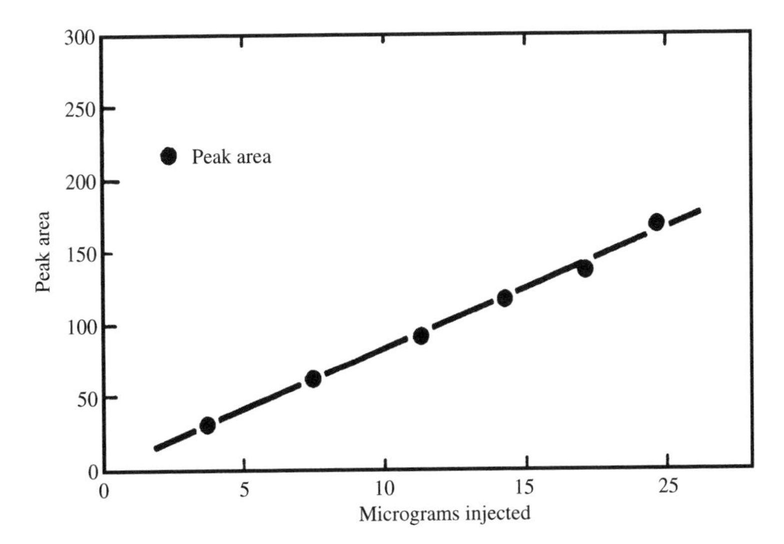

Figure 2.4 Peak area calibration with external standard

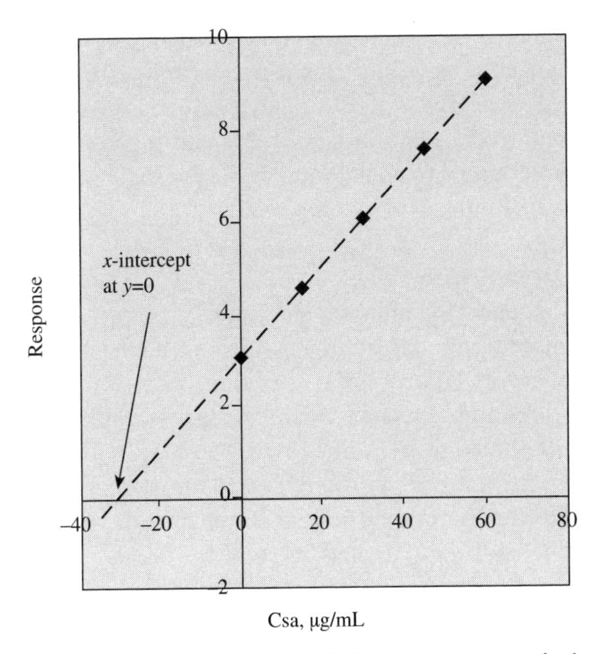

Figure 2.5 Example of the determination of the concentration of a known analyte by standard addition

and consequently the unknown concentration C_0 can be easily calculated as:

$$C_0 = -CV_f/V_u$$

In other words, by spiking a solution containing the analyte of interest at unknown concentration with known volumes of the analyte standard solution, it is possible to determine its concentration in the original solution.

2.1.4.3 *Internal Standard*

In general the analytical procedure is affected by various analytical errors, deriving from either sample and standard solutions manipulations or instrumental problems (e.g. instability of instrumental response). For these reasons the approach followed implies the addition of an internal standard (IS), which compensates for the various analytical errors and is particularly effective in the case of highly complex substrates. A known compound at a fixed concentration is added to the unknown sample before any sample treatment and purification procedure to give a separate peak in the chromatogram.

It follows that only a high structural similarity between the analyte of interest and the IS can guarantee an equal efficiency of extraction of both from the analytical sample and this aspect is fully satisfied when stable isotope labelled derivatives of the analyte are used as the IS. In these conditions a biunivocal relationship is obtained between the ratio of the peak areas corresponding to analyte and IS and the molar ratio [analyte]/[IS].

Three kinds of IS are usually employed:

(i) A compound with r.t. value different from that of the analyte, homologous to the analyte, leading to a fragment A^+ identical to the analyte fragment. In SIM conditions a chromatogram such as that shown in Figure 2.6(a) is obtained.
(ii) A compound of the same chemical class of the analyte leading to ions with the same *m/z* value [in this case, as in (i), it must exhibit a different r.t. value] or a different *m/z* value (in this case the r.t. value can be different or equal to that of the analyte) (see Figure 2.6b).
(iii) A stable isotope labelled derivative of the analyte (usually containing ^{2}H, ^{13}C, ^{15}N or ^{18}O). It will exhibit the same r.t. value as the analyte but different *m/z* value (Figure 2.6c). This represents the most valid approach, but it clashes with the cost of labelled standard compounds.

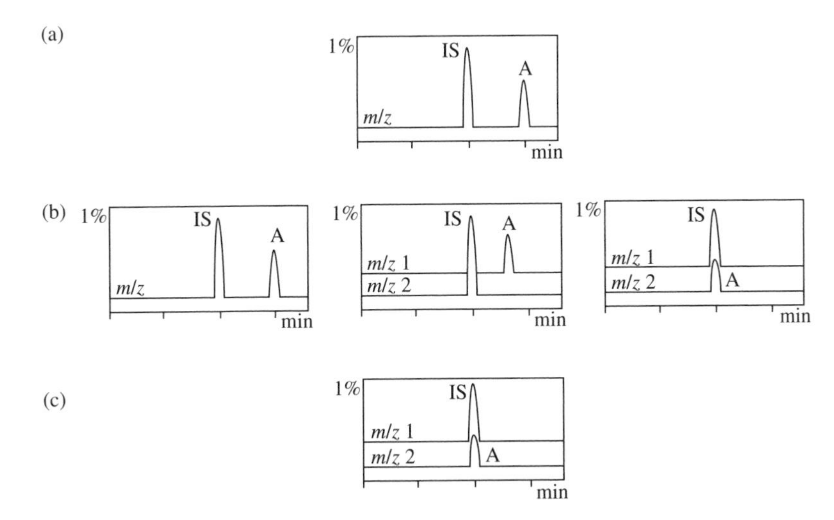

Figure 2.6 Examples of the different results obtained in SIM conditions by the use of different IS: (a) compound with a r.t. different from the analyte, leading to a fragment of the same *m/z* value of that chosen for the analyte (A) detection; (b) compound of the same chemical class of the analyte giving a ionic species isobaric with the analyte (left) or with a different *m/z* value and different (centre) and equal (right) retention; (c) a stable isotope labelled derivative of the analyte. It will exhibit the same r.t. of the analyte, but a different *m/z* value

It must be emphasized that labelled standard must satisfy some requirements. Thus, in the case of deuterated derivatives:

- at least three deuterium atoms must be present in the molecule, to lead to a mass shift of three units from the molecular species of the analyte. For a lower number of deuterium atoms, possible interferences with the isotopic peak of the analyte (M+1, natural ^{13}C-containing and M+2 natural ^{18}O-containing species) can occur;
- the percentage of deuterium must be as high as possible and the percentage of undeuterated component must be as low as possible;
- the deuterium atoms must be introduced in well defined positions of the molecule, without changing the stereochemistry of the molecule itself;
- the deuterium atoms must be inserted in stable positions, without the possibility of any exchange with hydrogen during storage and use.

As an example, the structure of $^{2}H_3$-testosterone employed as IS in testosterone quantitative analysis and the SIM chromatogram related to testosterone and $^{2}H_3$-testosterone are reported in Figure 2.7.

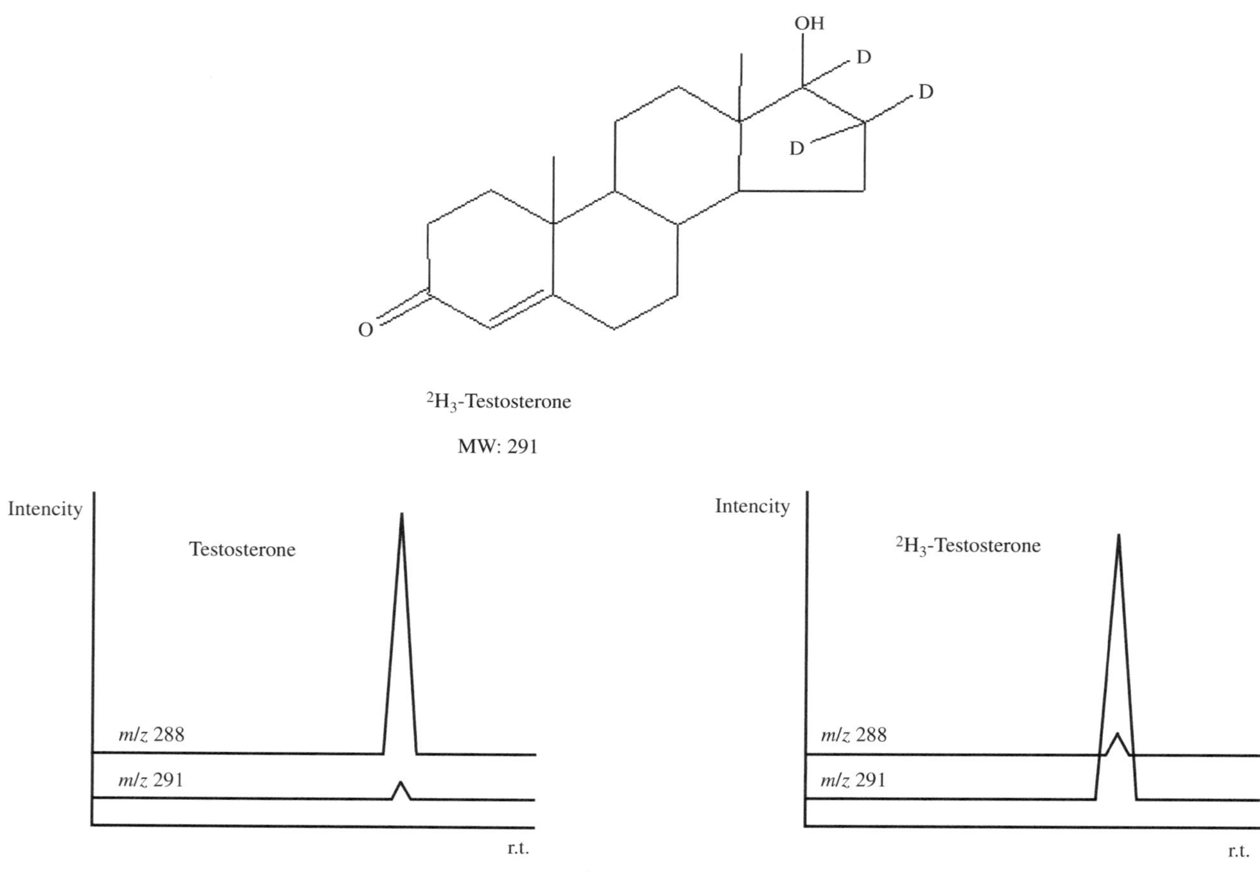

Figure 2.7 SIM chromatogram of testosterone (m/z 288) and $^2H_3$-testosterone (m/z 291)

Run	1	2	3	4	5
Testosterone (pg/mL)	0	2.5	25	100	250
$^2H_3$-testosterone (pg/mL)	100	100	100	100	100
[Testosterone]/[$^2H_3$-testosterone]	0	0.025	0.25	1.0	2.5

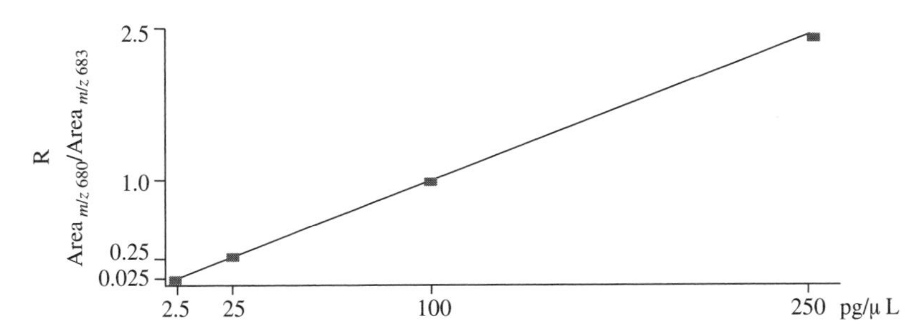

Figure 2.8 Testosterone, derivatized by HFBA, calibration curve

The calibration curve is obtained by analysis of different samples with different [analyte]/[IS] molar ratio, taking the concentration of IS as constant. The data obtained for the calibration curve of testosterone, derivatized by hexafluorobutyric acid (HFBA), are reported in Figure 2.8. The calibration curve has been obtained from the partial SIM chromatogram shown in Figure 2.9, from which the peak areas of analyte and IS can be easily calculated.

At this stage some comments on the use of stable isotope labelled derivatives as IS are required. Let us consider a theoretical SIM experiment, in which the analyte and the IS with masses different by more than three units are injected in quantities A and S, respectively. Consider that the analyte and the IS give rise to ions at *m/z* values m_1 and m_2, respectively. The contributions of the analyte to m_1 and m_2 intensities are a_1A and a_2A respectively, while the contributions of the IS to m_1 and m_2 are b_1S and b_2S. In these conditions, the ratio R of the measured peak areas is given by:

$$\mathrm{R} = \frac{a_1\mathrm{A} + b_1\mathrm{S}}{a_2\mathrm{A} + b_2\mathrm{S}} \tag{2.3}$$

If the isotope labelled IS is isotopically pure, the contribution of the analyte to mass m_2 is negligible, $a_2\mathrm{A} << b_2\mathrm{S}$, and Equation (2.3) becomes:

$$\mathrm{R} = (a_1/b_2\mathrm{S})\mathrm{A} + b_1/b_2 \tag{2.4}$$

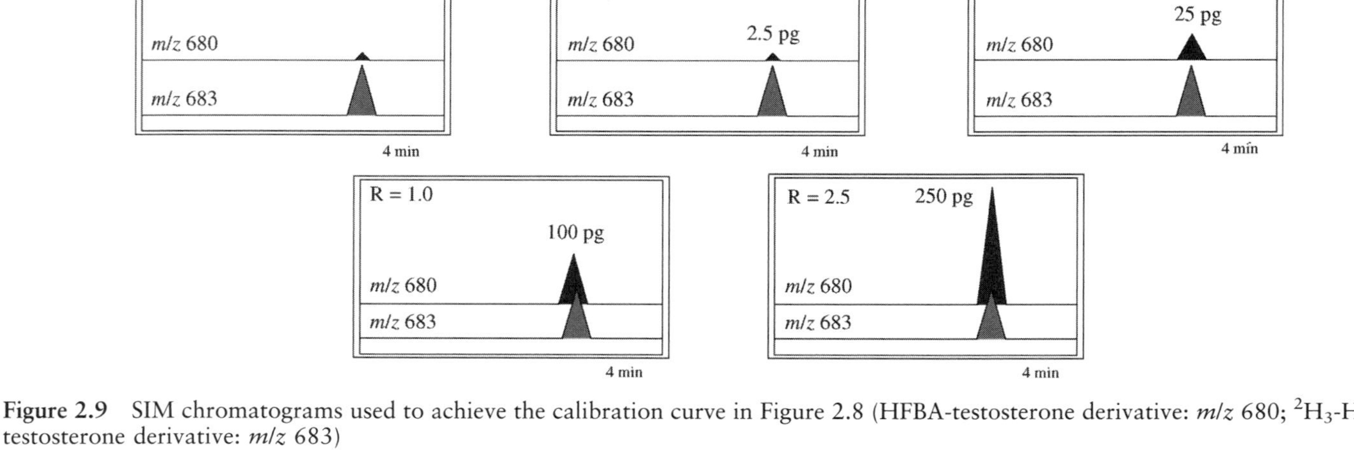

Figure 2.9 SIM chromatograms used to achieve the calibration curve in Figure 2.8 (HFBA-testosterone derivative: *m/z* 680; $^{2}H_{3}$-HFBA-testosterone derivative: *m/z* 683)

When the IS is not isotopically pure, b_2S becomes smaller, the use of a linear regression straight line as the calibration curve is incorrect and the calibration function can be only approximated by a quadratic equation.[1]

2.1.5 Method Validation

At the end of this chapter dealing with the strategy to perform a correct analysis and with the performance characteristics (see above) which allow its evaluation, a question arises whether the analytical procedure meets the intended requirements. The process by which the control is done is called 'method validation'. Even if there are different opinions concerning the terminology of the performance characteristics, method validation is generally defined as 'validation of a method establishes, by systematic laboratory studies, that the method is fit-for-purpose, i.e. its performance characteristics are capable of producing results in line with the needs of the analytical problem'.[2]

REFERENCES

1. J. R. Chapman and E. Bailey, *J. Chromatogr.*, **89**, 215 (1974).
2. CITAC/EURACHEM, *Guide to Quality in Analytical Chemistry: An Aid to Accreditation* (2002).

3

How to Improve Specificity

Of course, any analytical method must be as specific as possible. This means that instrumental data taken into consideration must be strictly related to the physico-chemical properties of the analyte of interest.

In this respect MS is a highly effective technique: for example, in most cases, the EI spectra well characterize, from a structural point of view, the organic molecule under study on the basis of molecular and fragment ions. This is generally (but not always) true when we consider a pure analyte: when it is present, at low concentration, in a complex mixture, its direct identification (and quantification) can be inferred by the presence of isobaric ions originating from other different compounds present in the mixture.

For these reasons the coupling with chromatographic [GC,[1] LC,[1,2] capillary zone electrophoresis (CZE),[3] gel permeation chromatography (GPC),[2] supercritical fluid chromatography (SFC)[4]] methods has been, and is considered, of high interest allowing on the one hand the achievement of mass spectral data of the analyte, possibly without any interfering species, and on the other giving further analytical data, i.e. the r.t. values of the different species present in the mixture.

However, even using a chromatographic/MS approach, sometimes the selectivity of the system is not enough to obtain valid and well reproducible measurements. Consider, for example, an analyte present at trace level in a matrix so complex to lead to a practically unresolved chromatogram. In this case the r.t. datum is lost. Furthermore, due to the very low level of the analyte, both RIC or multiple ion monitoring measurements may fail, leading to signals with a very low signal-to-noise ratio, due to the presence of co-eluting species leading to isobaric molecular or fragment ions.

Quantitative Applications of Mass Spectrometry I. Lavagnini, F. Magno, R. Seraglia and P. Traldi

In these cases a higher specificity is required; this chapter is devoted to the possible instrumental approaches which can be employed for this aim.

3.1 CHOICE OF A SUITABLE CHROMATOGRAPHIC PROCEDURE

Of course this choice is strictly related to the physico-chemical properties of the analyte of interest. However, it can be also influenced by two other aspects:

(i) the instrumental availability in the laboratory;
(ii) the 'cultural' tradition of the researchers operating in the laboratory.

In other words, this choice is affected by the instruments present in the laboratory and by the knowledge and past experience of the researchers. What is often observed in the literature is that different chromatographic/MS approaches are proposed for the same analyte.

In this chapter the methods most widely employed, i.e. GC/MS, and LC/MS, will be described and some examples will be given.

3.1.1 GC/MS Measurements in Low and High Resolution Conditions

A chromatographic process is the result of different phenomena:

- thermodynamic (due to analyte–mobile phase–stationary phase interactions)
- kinetic (due to relative motion of the analyte).

A GC system is shown schematically in Figure 3.1. The sample is vaporized into a heated injector (sample introduction in Figure 3.1) and is transferred inside the column by the action of the carrier gas (mobile phase). Separation takes place by the interaction of the various analytes with the stationary phase.[1]

Two different kinds of columns are usually employed for chromatographic separation: packed and capillary ones, with internal diameter >0.5 and <0.5 mm, respectively. The former exhibit higher retention

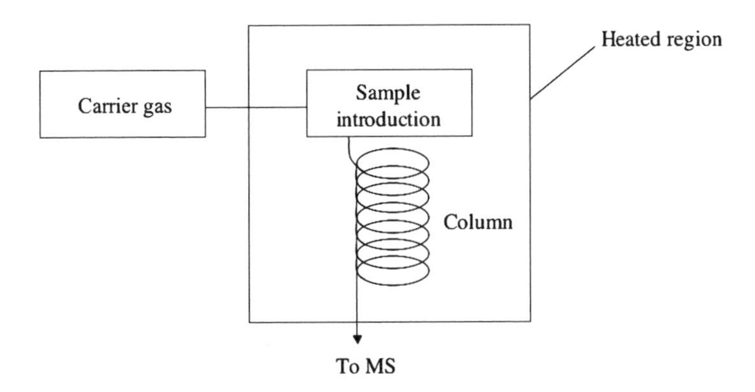

Figure 3.1 Scheme of a gas chromatographic system

volume and capacity, while the latter show higher efficiency and permeability.

Due to its performance, as well as to the lower carrier gas flow, allowing direct coupling with a mass spectrometer, the latter approach is usually employed.

The coupling of GC with MS leads to positive aspects for both techniques:

- from the MS point of view, the coupling allows the introduction inside the ion source of the different analytes, (possibly) well separated from each other;
- from the GC point of view, MS is a sensitive, universal and specific detector.

Hence, an increase in specificity is innate in the GC/MS approach, in particular when the SIM and MID methodologies described in Chapter 1 are employed. However, a further, and noteworthy, increase can be obtained when the mass spectrometer is operating at high resolution conditions. As described in Chapter 1, high resolution allows the accurate mass value of ions generated from different analytes to be determined. The accurate mass is dependent on the elemental composition of the ions. Hence, isobaric ions (i.e. exhibiting the same integer mass) of different elemental composition exhibit different values of exact mass. These different values can be evidenced only by high resolution measurements.

This aspect can be validly employed to increase the selectivity of a measurement. Consider, as an example, the partial chromatograms

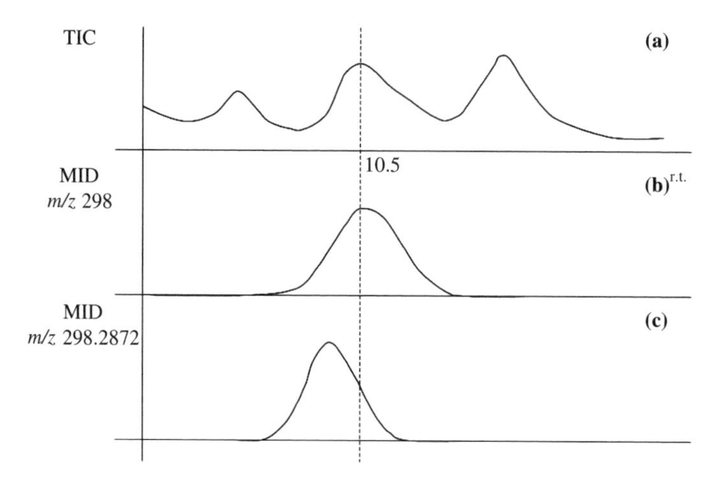

Figure 3.2 Partial gas chromatogram of a mixture containing methyl stearate obtained in different operating conditions: (a) by TIC; (b) by low resolution multiple ion detection of $M^{+\cdot}$ of methyl stearate (*m/z* 298); (c) by high resolution (10 000) multiple ion detection of $M^{+\cdot}$ of methyl stearate (*m/z* 298.2872)

reported in Figure 3.2. The TIC chromatogram (Figure 3.2a) shows, in the r.t. window of interest, a series of peaks. Consider that we are interested in the quantification of methyl stearate. Its elemental formula, $C_{19}H_{38}O_2$, leads to an accurate mass value of 298.2872. If we operate in low resolution conditions, we can perform a SIM experiment selecting the integer mass value of 298. Using this approach the chromatogram of Figure 3.2(b) is obtained, which indicates that the peak at r.t. 10.5 min corresponds to species producing ions (molecular or fragment) at *m/z* 298. At first sight this MID peak could be employed for methyl stearate quantification but, when selectivity is increased by increasing the MS resolution, it follows that only one portion of the chromatographic peak at 10.5 min is due to methyl stearate (Figure 3.2c). The other portion of the peak is due to another ionic species with the same integer *m/z* value (298), but with a different accurate mass value. These two species were unresolved when operating in low resolution mode and consequently quantitative measurements performed in these conditions would lead to incorrect results.

This example does not mean that high resolution conditions are always essential for the achievement of valid GC/MS results. In most cases, low resolution MS is more effective and an example of the validity of this approach will be given in the next section. However, in some other cases, in the particular in the presence of highly complex matrices and with analytes at levels lower than ppb, the use of more specific and selective methods are required; an example will be given in Section 3.3.

3.1.1.1 *An Example of GC/MS Application of Biomedical Interest: Quantitative Determination of Glyoxal and Methylglyoxal in Plasma Samples*[5]

Determination of glyoxal and methylglyoxal levels in plasma is of great interest, because it gives information on the oxidation processes occurring in glycated protein. A simple method based on GC/MS measurements, has been recently developed. Considering the high complexity of plasma, a sample treatment, summarized in Figure 3.3, is required.

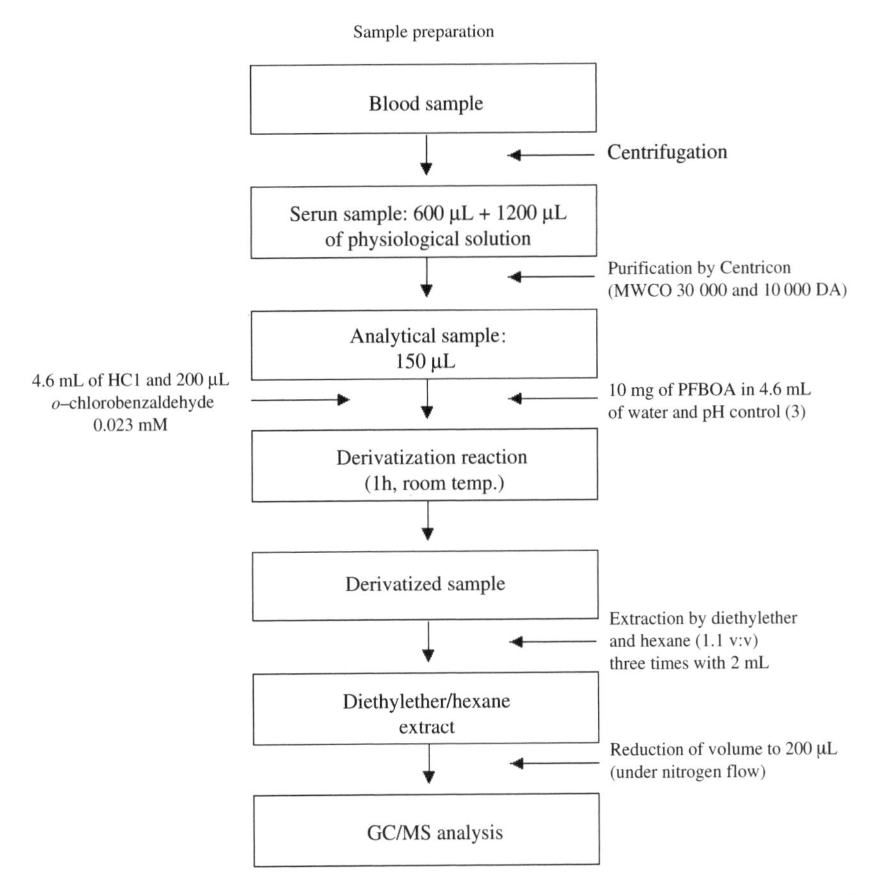

Figure 3.3 The measurement of glyoxal and methylglyoxal in plasma. Sample treatment procedure before GC/MS analysis

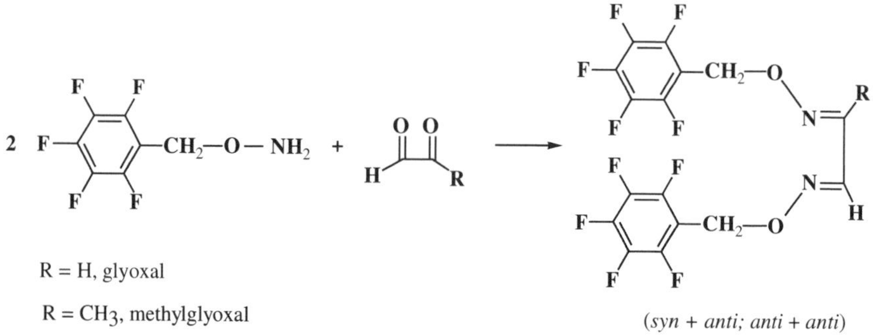

R = H, glyoxal

R = CH_3, methylglyoxal

(*syn* + *anti*; *anti* + *anti*)

Figure 3.4 Derivatization reaction of glyoxal and methylglyoxal with PFBOA before GC/MS analysis. Reproduced from Lapolla *et al.* (2003)[5], with permission from John Wiley & Sons, Ltd

Glyoxal and methylglyoxal were evaluated after derivatization with O-(3,3,4,5,6-pentafluorobenzyl)hydroxylamine hydrochloride (PFBOA). The derivatization procedure, described in the literature for glyoxal and methylglyoxal measurements in various natural matrices[6] and partially modified for this purpose,[5] leads to the formation of O-PFB oximes, as shown in Figure 3.4. To obtain the calibration curve, four solutions of glyoxal with different concentrations (5.5, 11, 55 and 166 μg/L) were prepared. In parallel, an IS (*o*-chlorobenzaldehyde) solution (0.023 mM) was prepared; 150 μL of the various glyoxal solutions and 200 μL of the IS were dissolved in 4.6 mL of an HCl solution (pH 3). After mixing, the pH was carefully checked and adjusted. The resulting solution was treated with 10 mg of PFBOA and stirred at room temperature for 1 h. After the addition of 0.5 mg of NaCl, O-PFB oximes were extracted with diethyl ether/hexane 1:1 v/v (3×2 mL), stirring for 5 min, and the organic phase was dried over Na_2SO_4 and filtered. The volume was reduced to 0.2 mL under nitrogen flow before GC/MS analysis. The mass spectrometer was operating in low resolution MID mode and signals at m/z 181 (base peak of PFBOA derivatives), 300 (IS) and 448 (the oximes derived from glyoxal) were recorded. The resulting calibration curve ($y = 4.463x + 16.009$) showed a high R^2 value (0.9978).

For the analysis of plasma samples, 4.6 mL of HCl solution (pH 3) were added to 150 μL aliquots of the final filtrate. To this solution, 200 μL of *o*-chlorobenzaldehyde solution (0.023 mM) were added and the pH was adjusted to 3 with a few drops of 0.1 M HCl. The derivatization procedure was exactly the same as that used for the calibration curve.

Table 3.1 Retention time of *syn* and *anti* *o*-chlorobenzaldehyde O-PFB oximes and *syn* + *anti* and *anti* + *anti* methylglyoxal and glyoxal O-PFB oximes

Peak	Compound	Retention time (min)
1	*o*-Chlorobenzaldehyde (IS)	31.53 ; 32.01
2	Methylglyoxal	31.79 ; 32.20
3	Glyoxal	32.73 ; 32.92

Methylglyoxal was identified according to literature data: oxime peak relative r.t. and monitoring the ions at *m/z* 181 and 462.[7,8] Compounds were quantified on the basis of the sum of signals; concentration of glyoxal and methylglyoxal were calculated on the calibration line of the sum of glyoxal *syn* and *anti* PFBOA dioxime signals. The r.t. of *syn* and *anti* *o*-chlorobenzaldehyde O-PFB oximes and *syn* + *anti* and *anti* + *anti* methylglyoxal and glyoxal O-PFB oximes are listed in Table 3.1.

To evaluate the effect of sample storage on glyoxal and methylglyoxal quantitative measurements, plasma samples were stored at −70 °C for 1 month: the glyoxal and methyglyoxal levels before and after storage remained practically the same, with variation remaining within the experimental standard deviation.

An example of the results achievable by this method is shown in Figure 3.5, illustrating the GC/MS results for a healthy subject and a diabetic patient (in the latter case, higher glyoxal and methylglyoxal are to be expected). Peak 1 is due to *o*-chlorobenzaldehyde, employed as IS, peaks 2 to the stereoisomeric O-PFB oximes of methylglyoxal and peaks 3 to the O-PFB oximes of glyoxal. As may be seen from the MID chromatogram of Figure 3.5, higher abundances of glyoxal and methylglyoxal were observed in the plasma sample from the diabetic patient.

The method was applied to the study of glyoxal and methylglyoxal levels for 10 poorly controlled type 2 diabetic patients and a control group of 20 subjects. The obtained results showed that the glyoxal and methylglyoxal mean levels were 27.2 ± 8.6 and 29.3 ± 5.5 µg/mL, respectively, for the former group, while in the case of healthy subjects these levels were 12.5 ± 0.5 and 8.5 ± 0.5 µglmL.

3.1.2 LC/ESI/MS and LC/APCI/MS Measurements

The development of a LC/MS analytical method depends on different factors, which will be briefly described.

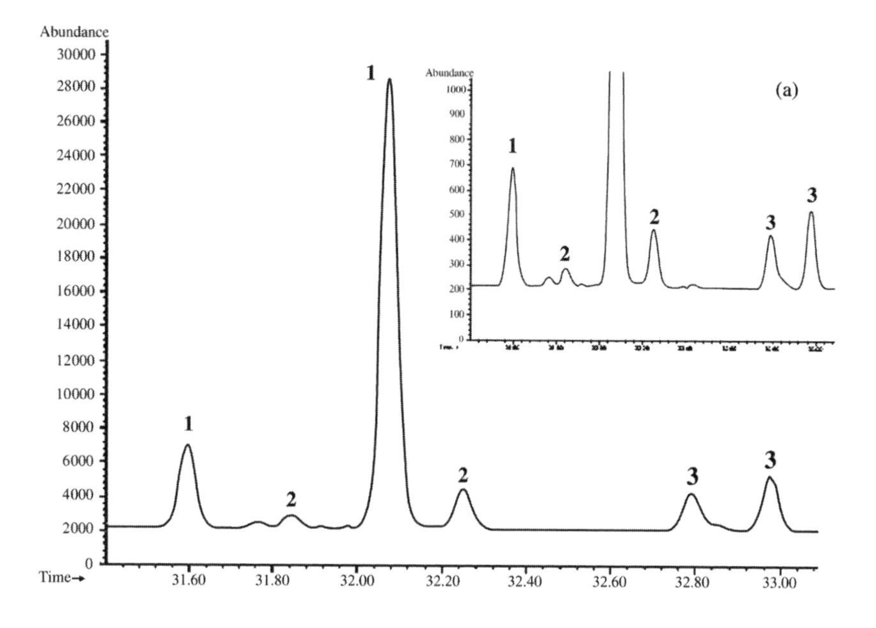

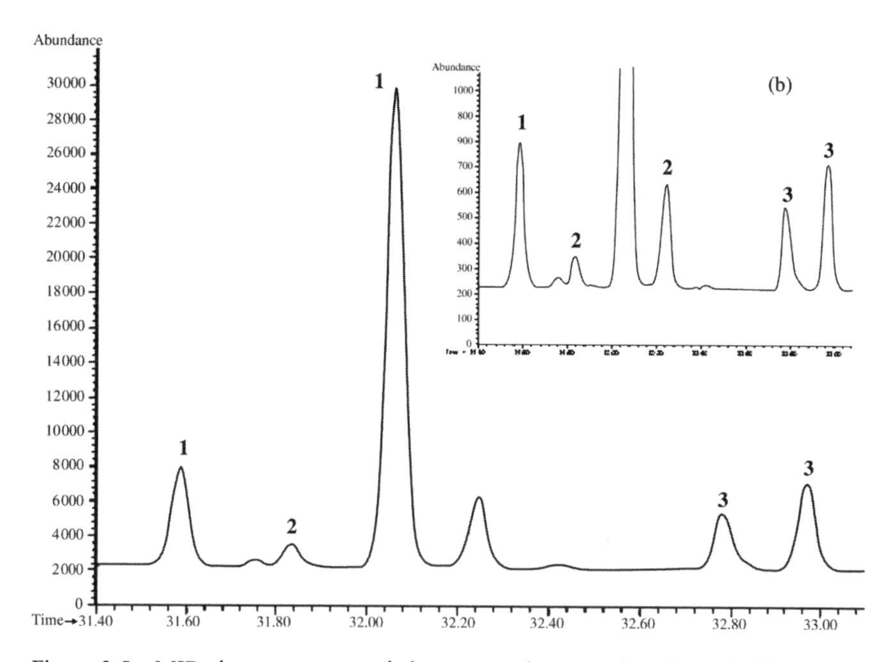

Figure 3.5 MID chromatograms of plasma samples treated as shown in Figure 3.3. (a) Diabetic patient in poor metabolic control; (b) healthy subject. Peaks 1 are due to *syn* and *anti o*-chlorobenzaldehyde O-PFB oximes employed as internal standard. Peaks 2 are due to the *syn* + *anti* and *anti* + *anti* methylglyoxal O-PFB oximes. Peaks 3 are due to the *syn* + anti and *anti* + *anti* glyoxal O-PFB oximes. Reproduced from Lapolla *et al.* (2003)[5], with permission from John Wiley & Sons, Ltd

First of all, the choice of the ionization technique strongly depends on the chemical properties of the analyte(s) of interest. In fact, as briefly described in Chapter 1, ESI and APCI are not 'universal' ionization techniques, due to the different ionization mechanisms operating in these API sources.

In general ESI is more compatible with the following analytes:

- ionic molecules;
- compounds containing heteroatoms (able to be easily protonated in solution);
- acidic and basic organic compounds;
- proteins, peptides and oligonucleotides (with formation of multiple charged ions).

APCI is more compatible with small molecules, which can be moderately polar to nonpolar and samples not necessarily containing heteroatoms [e.g. polyaromatic hydrocarbons, polychlorobiphenyls (PCBs), fatty acids, triglycerides, benzodiazepines, carbamates], while thermally (or photosensitive) compounds are not suitable for this approach: in fact they may decompose or react in the heated nebulizer region of the ion source. Moreover, APCI/MS analysis of proteins, peptides and oligonucleotides should be avoided, due to a severe decrease in sensitivity with respect to ESI/MS.

A relevant parameter to be considered in the set-up of a LC/MS analytical procedure is the identification of the proper LC method. Also the different ionization mechanisms operating in the two API sources must be considered. The LC methods compatibility with ESI and APCI are summarized in Table 3.2.

Once the most convenient LC method and the appropriate API technique to be employed are identified, particular attention must be given to the eluent flow rate and composition.

Table 3.2 LC method compatibility with API sources

LC mode	ESI	APCI
Reversed phase	High	Moderate
Normal phase	Low	High
Size exclusion	High	Low
Ion pair	Moderate	Moderate
Ion exchange	Low	Low
Hydrophobic interaction	Low	Low
Immunoaffinity	High	Low

Table 3.3 Flow rate and API interface compatibility

Column ID (mm)	Typical flow rate (mL/min)	Compatibility	
		ESI	APCI
4.6	1.0	Yes (split 10:1)	Yes
3.9	0.5	Yes (split 5:1)	Yes
2.1	0.2	Yes (split 1:1)	Yes
1.0	40–50 μL/min	Yes	No
Capillary	<10 μL/min	Yes	No

The eluent flow rate compatibility of ESI and APCI are different. In fact, while "classical" ESI can accept flow rate in the range of 1–5 μL/min to 1 mL/min, APCI works appropriately with flow rate in the range 0.5–4 mL/min. These differences can be ascribed to the different ionization mechanisms operating in these two API sources.

In LC/APCI/MS analysis, an irreproducible analyte signal is observed operating with flow rate below 0.4 mL/min due to the formation of an unstable plasma of ionized solvent molecules around the discharge needle, which does not lead to the proper ionization of analyte(s). In contrast, at high eluent flow rate the droplets formed in the ESI source are large and they take longer to reach the point of fission. Consequently, the process of ion production in the atmospheric pressure source (region) may not be optimal. The typical flow rates of the different LC columns together with the compatibility with ESI and APCI are reported in Table 3.3. As can be seen, the column internal diameter should be chosen to best fit the optimal flow rate range of the interface employed (Figure 3.6). Depending on the geometry and types of ESI source available in the laboratory, the use of flow splitting procedure may be necessary for optimizing the instrumental response of analyte(s). Henion and Maylin[9] have shown that a detection limit of 500 pg for methylparathion can be obtained using a 4.6 mm ID column with a split ratio of 99:1, while, using a 0.5 mm ID, fused silica packed column without splitter, 50 pg of phenotiazine can be easily detected.

A negative aspect of the use of a column with flow rate <0.1 mL/min is the need for special LC pumping systems, required to maintain the flow rate, to reduce the gradient dwell times and to ensure low dead volumes.

Regarding the eluent composition, many solvents, commonly used in traditional LC analysis, are compatible with API interfaces.

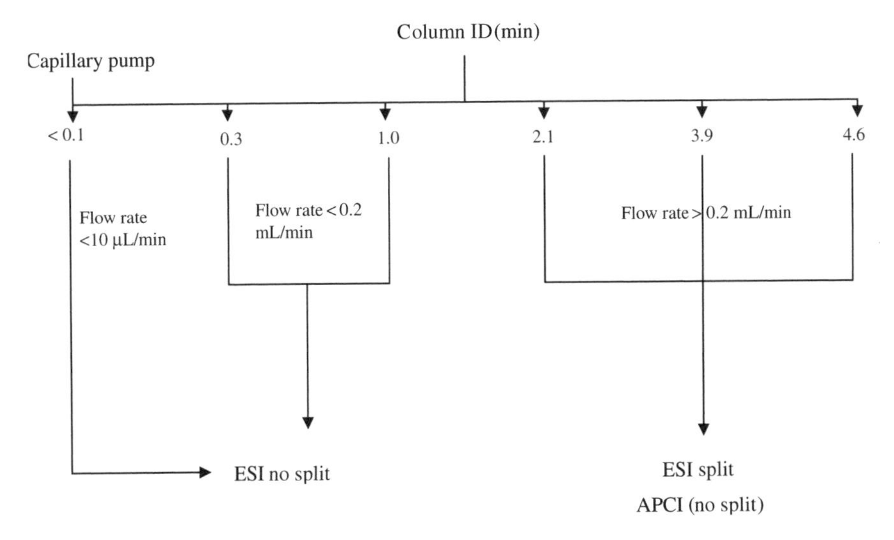

Figure 3.6 Relationship between LC column internal diameters, flow rates and ESI/APCI operating conditions

Using an ESI source, solvents with low surface tension, high conductivity and low heat of solvation are preferred. The optimization of these parameters leads to a correct droplet nebulization and desolvation with a high yield of ion production. However, even if water is a good solvent for ions, its high surface tension makes desolvation and ion desorption quite difficult. As can be seen from Figure 3.7, decreasing the water percentage in the solvent used, an increase in the analyte signal is obtained. In general, high percentages of organic modifiers in the LC eluent lead to an increase in sensitivity. This is particularly evident when a gradient LC/MS run is employed: the absolute analyte response can change when comparing early and later eluting species, as shown in Figure 3.8. Figure 3.8 shows the chromatogram of a solution containing homologous alcohols at the same concentration obtained by gradient

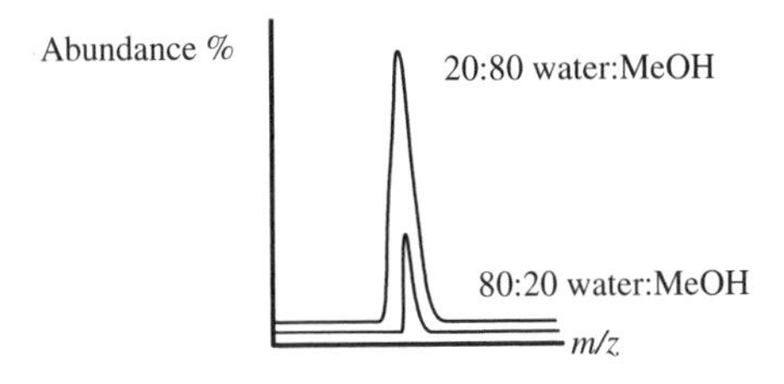

Figure 3.7 LC/ESI/MS analysis of penicillinG solutions in water:methanol 20:80 (v:v) and 80:20 (v:v)

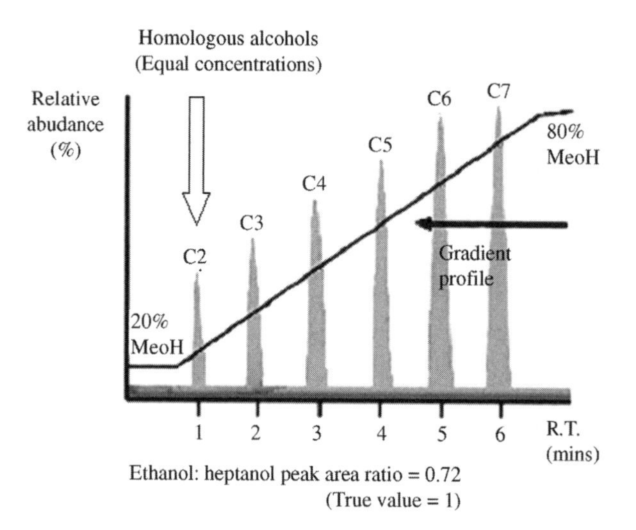

Figure 3.8 Chromatogram of an equimolar solution of homologous C2–C7 aliphatic alcohols obtained by gradient LC/ESI/MS analysis

LC/ESI/MS analysis. As can be seen, the response of heptanol (r.t. about 6 min) is higher than that of ethanol (r.t. about 1 min), due to the higher percentage of methanol present in the eluent. In other words, to obtain valid results by quantitative analysis by gradient LC/ESI/MS, the r.t. of the IS should be very close to that of the analyte(s) of interest.

In LC/APCI/MS, the eluent composition is not as critical as in LC/ESI/MS. Taking into account that proton transfer reaction is the most common ionization mechanism active in APCI conditions, the relative proton affinity of the solvent and of the analyte must be balanced for optimizing the sensitivity. Generally, methanol is a better choice than acetonitrile in LC/APCI/MS measurements.

For both the ionization techniques particular attention must be paid to the choice of the buffer. Volatile buffer [i.e. ammonium acetate, ammonium formate, acetic acid, formic acid (FA) or trifluoroacetic acid (TFA)] is preferred to nonvolatile buffer (i.e. phosphates, sulfates, borates and citrates), because the latter can reduce the sensitivity and lead to signal suppression, as show in Figure 3.9. Buffers must be chosen so that they have low molecular weight, high charge density and they must be hydrophilic.

In general, ESI interface can tolerate volatile buffer concentration less than 20 mM, while using an APCI source the volatile buffer concentration can be up to 100 mM. For nonvolatile buffer, the concentration must be kept as low as possible to avoid ion source contamination.

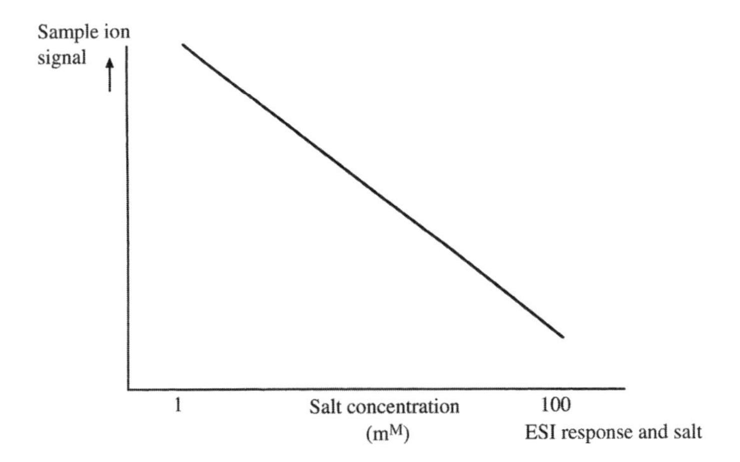

Figure 3.9 Relationship between LC/ESI/MS signal intensity and salt concentration: for high salt concentration signal suppression phenomena become relevant

The choice of buffer concentration is a compromise between good chromatographic separation and good MS response.

As stated above, ESI is highly effective for the analysis of peptides, but their quantitation in complex biomatrices at low concentration level (in the order of μg/mL or ng/mL) is a difficult task, owing to interference from the proteins and peptides present in high amounts in biological samples. A simple and rapid method has been recently developed for the determination of a novel antiviral peptide, sifuvirtide, in monkey plasma.[10] The amino acid sequence of this peptide is reported in Figure 3.10(a). The analytical method employed was based on the coupling, via a timed valve switching, of a SPE column with a LC/ESI/MS instrument, operating in MS/MS mode. The IS used was a synthetic analogue peptide, $^{127}I_4$-sifuvirtide (Figure 3.10b), containing four ^{127}I atoms on the tyrosine residues of the peptide and showing the same MS behaviour and physico-chemical characteristics of the unlabelled peptide. For both peptides, the ESI/MS positive ion mass spectra, obtained by infusing a 10 μg/mL solution of sifuvirtide and its $^{127}I_4$-derivative in water/acetonitrile (50:50; v/v) at a flow rate of 5 μL/min, are characterized by an intense peak corresponding to $[M+5H]^{5+}$ ions at *m/z* 946.5 and 1047.2, respectively, together with less intense peaks corresponding to $[M+4H]^{4+}$ and $[M+3H]^{3+}$. Once the MS behaviour of the peptides was characterized, better conditions for LC analysis were tested. Taking into account that the addition of an organic acid, such as FA or TFA, into the mobile phase generally leads to a better LC separation and to a

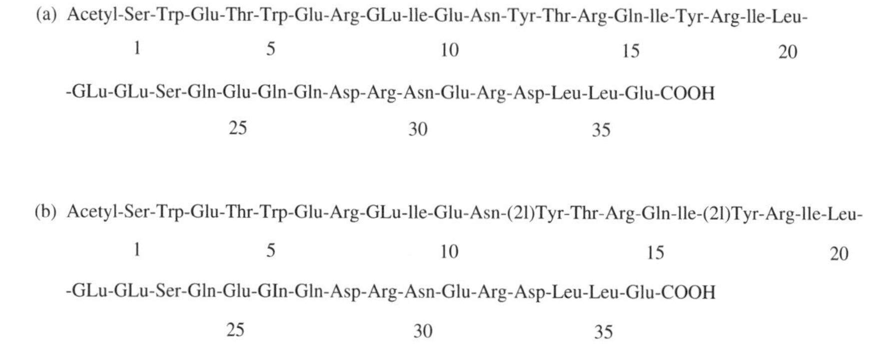

Figure 3.10 Amino acid sequences of (a) sifuvirtide and (b) $^{127}I_4$-sifuvirtide (employed as IS). Reproduced from Dai *et al.* (2005)[10] with permission from John Wiley & Sons, Ltd

better ionization yield in ESI conditions, different percentages of FA and TFA (ranging from 0 to 5 %) were added to the mobile phases. The mobile phases considered were water/acetonitrile and water/methanol 50/50 v:v. Better results were achieved using water/acetonitrile with 0.1 % FA, even if no significant differences were found using acetonitrile or methanol as organic modifiers. In order to obtain higher specificity and selectivity, the analysis were performed in MS/MS conditions, by monitoring collisionally generated fragments of the analyte and of the IS. For both peptides, a relatively high abundance of $[M+5]^{5+}$ ions was obtained and used for MS/MS experiments. The LC/ESI/MS/MS spectra of $[M+5]^{5+}$ ions of both peptides are reported in Figure 3.11, where intense peaks originating from $(b_{33})^{5+}$ peptide fragments are abundant for both compounds. The selected reaction monitoring (SRM) detection set up for quantitative analysis of sifuvirtide was the transition *m/z* 946.5 ($[M+5]^{5+}$) $\rightarrow$ 871.8 [$(b_{33})^{5+}$] for the analyte and for IS, $^{127}I_4$-sifuvirtide, was *m/z* 1087.2 ($[M+5]^{5+}$) $\rightarrow$ 972.3 [$(b_{33})^{5+}$]. Calibration samples were prepared by spiking known quantities of the sifuvirtide into deactivated drug-free plasma. Each calibration curve consisted of 100 μL of blank samples (plasma sample processed without IS), 100 μL of zero samples [plasma sample processed with 10 μL of IS working solution (5 μg/mL)], and 100 μL of six samples with a concentration of sifuvirtide ranging from 4.88 to 5000 ng/mL [processed with 10 μL of IS working solution (5 μg/mL)]. Then 100 μL of analyte and IS solution were introduced onto the extraction column for on-line preparation and LC/MS/MS analysis. The chromatograms of SRM analysis of sifuvirtide in plasma are shown in Figure 3.12, while the related calibration curve is

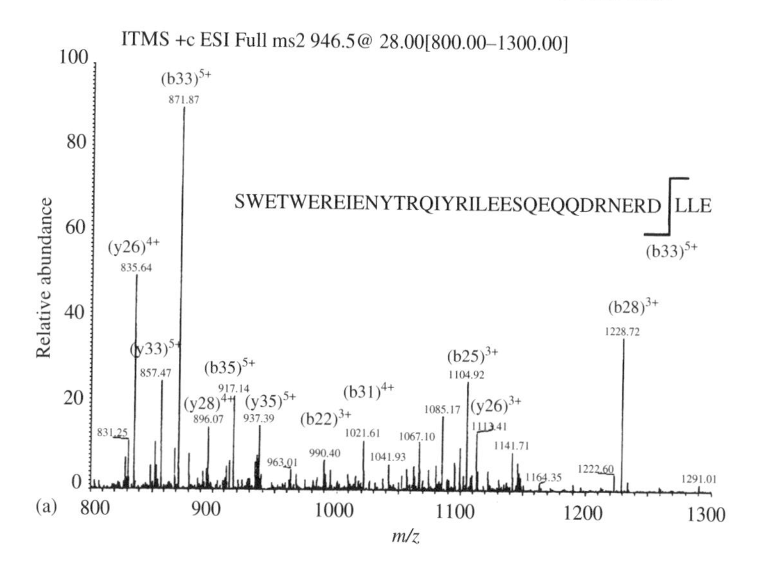

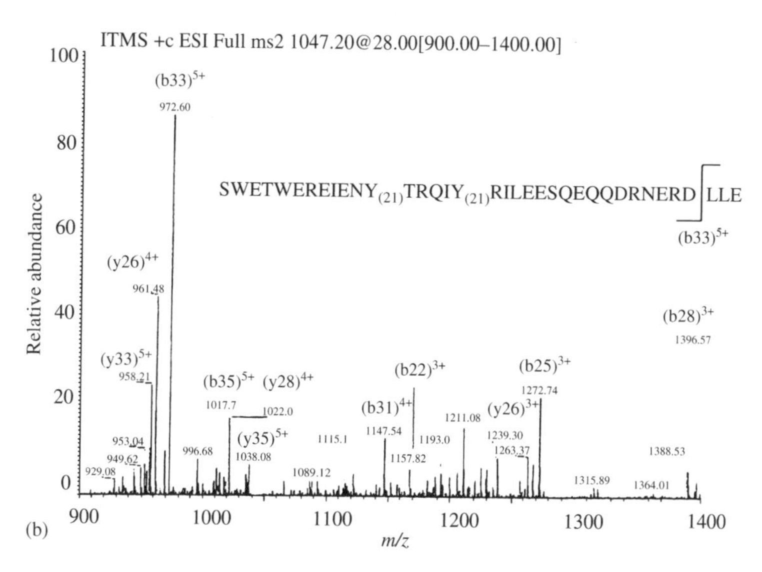

Figure 3.11 Collisional spectra (MS/MS) of the species: (a) $[M+5]^{5+}$ (*m/z* 946) for sifuvirtide; (b) $[M+5]^{5+}$ (*m/z* 1047) for the IS. Reproduced from Dai *et al.* (2005)[10] with permission from John Wiley & Sons, Ltd

reported in Figure 3.13. Validation of the method demonstrated that the linear calibration curve covered the range 4.88–5000 ng/mL, and the correlation coefficients were above 0.9923. The limit of detection (LOD) with signal-to-noise ratio higher than 12 was calculated as 1.22 ng/mL. The intra- and inter batch precision were less than 12.7 and 9.1 %,

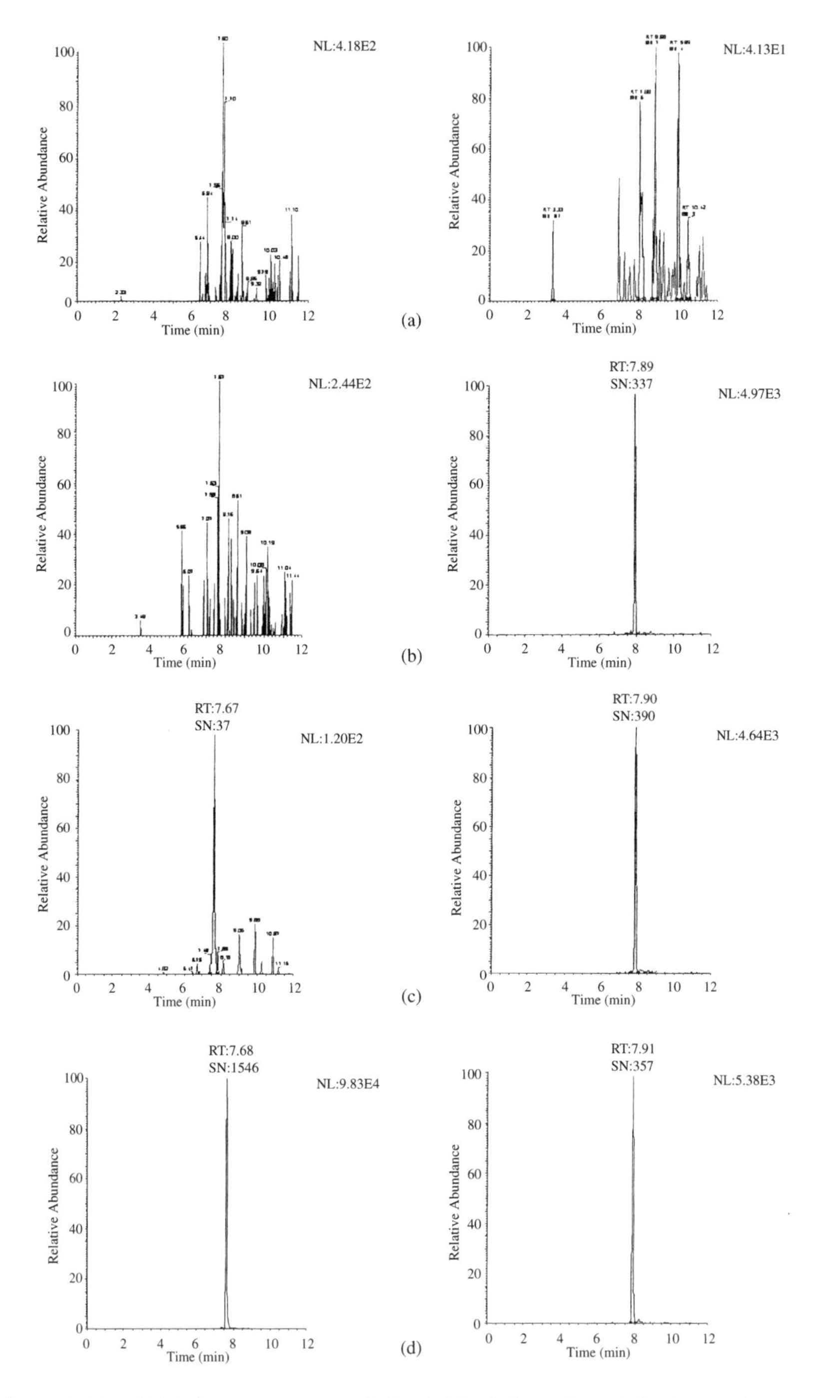

Figure 3.12 SRM chromatograms of sifuvirtide (left) and IS (right) in: (a) drug free monkey plasma; (b) at zero concentration but with the addition of IS; (c) lower limit of quantitation (LOQ) (4.88 ng/mL); (d) upper LOQ (5000 ng/mL). Reproduced from Dai *et al.* (2005)[10] with permission from John Wiley & Sons, Ltd

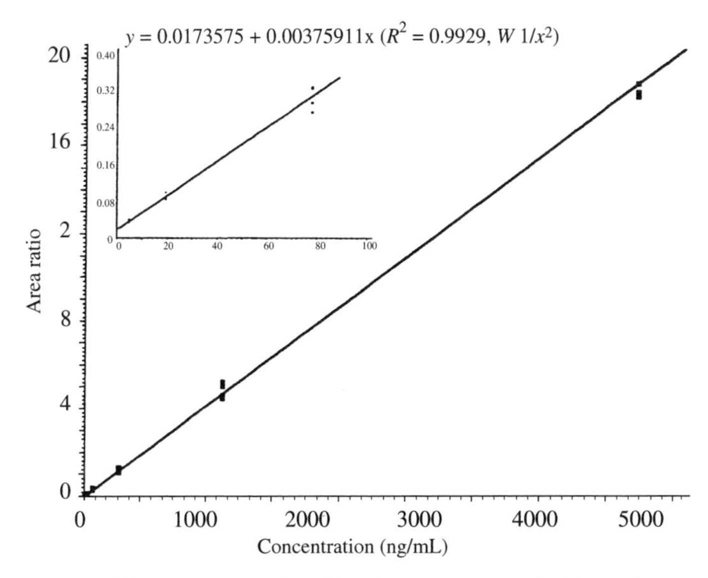

Figure 3.13 Calibration curve for the determination of sifuvirtide in monkey plasma using $^{127}I_4$-sifuvirtide as IS. Reproduced from Dai *et al.* (2005)[10] with permission from John Wiley & Sons, Ltd

and the mean accuracy ranged from −5.2 to 3.6 %, respectively (see Table 3.4). This analytical method has been successfully applied for the determination of a pharmacokinetic profile in preclinical studies on the administration of this new antiviral peptide.

The use of LC/MS technique in clinical and forensic toxicology is growing due to the possibility of obtaining fast drug assay with high sensitivity and selectivity. An example is the LC/APCI/MS procedure applied for a rapid and reliable screening, identification and quantitation of 23 benzodiazepines (alprazolam, bromazepam, brotizolam, camazepam, chlordiazepoxide, clobazam, clonazepam, diazepam, flunitrazepam, flurazepam, desalkylflurazepam, lorazepam, lormetazepam, medazepam, metaclazepam, midazolam, nitrazepam, nordazepam, oxzepam, prazepam, temazepam, tetrazepam, triazolam), their antagonist flumazenil and the benzodiazepine BZ_1 (omega 1) receptor agonist zoleflone, zolpidem and zopiclone in plasma from subtherapeutic to overdose concentration.[11] The aim of this work was to develop a multidrug assay for these classes of compounds, for which methods for the determination of only single substances, or a mixture of a few of them, were published. The analytes were purified from plasma using liquid–liquid extraction and separate by LC using gradient elution with aqueous ammonium formate (pH 3) and acetonitrile. An APCI source

Table 3.4 Accuracy and precision of sifuvirtide determination in pooled plasma from healthy rhesus monkeys at different concentration levels

	Concentration (ng/mL)						
Group	**4.88**	**19.5**	**78.1**	**312.5**	**1250**	**5000**	**Linearity**
Day 1	5.5	19.5	83	342	1209	4853	$y = A + Bx$
	4.3	17.4	71	324	1313	5153	$A = 0.03748$
	5.1	18.6	74	318	1310	4904	$B = 0.002107$
Mean	4.9	18	76	328	1277	4970	$R^2 = 0.9939$
SD	0.6	1.1	6	12	59	161	
RSD(%)	12.7	5.8	7.8	3.8	4.6	3.2	
Accuracy(%)	1.3	−5.4	−2.7	5.0	2.2	−0.6	
Day 2	4.6	18.7	77	301	1229	5235	$y = A + Bx$
	5.5	17.4	71	323	1275	5078	$A = 0.01846$
	4.8	18.6	86	296	1391	5124	$B = 0.003611$
Mean	5.0	18.2	78	307	1298	5146	$R^2 = 0.9945$
SD	0.4	0.8	7	14	83	81	
RSD(%)	9.1	4.2	9.5	4.6	6.4	1.6	
Accuracy(%)	1.6	−6.6	−0.2	−1.8	3.9	2.9	
Day 3	5.1	20.9	84	275	1218	4832	$y = A + Bx$
	4.4	17.8	81	274	1402	5103	$A = 0.01487$
	5.2	17.7	83	313	1311	5111	$B = 0.003985$
Mean	4.9	18.8	83	287	1310	5015	$R^2 = 0.9923$
SD	0.5	1.8	1.9	22	92	159	
RSD(%)	9.3	9.8	2.3	7.8	7.1	3.2	
Accuracy(%)	0.6	−3.6	5.9	−8.1	4.8	0.3	
3-day mean	4.9	18.5	79	307	1295	5044	
3-day SD	0.5	1.2	5.7	23.0	70.3	143.6	
Total RSD(%)	9.1	6.3	7.2	7.5	5.4	2.8	
Total accuracy(%)	1.2	−5.2	1.0	−1.6	3.6	0.9	

SD, standard deviation; RSD, relative standard deviation.

was preferred to an ESI one because it is much less susceptible to ion suppression.[12,13] After screening and identification in full-scan mode of all the analytes, their quantitation was performed by MID analysis. For quantitative analysis the analytes were divided into four different groups according to their therapeutic range and each group was assigned to one of four separately recorded traces with specific gain values as given in Table 3.5. The target ions of compounds eluting close to the end of a time window were monitored also in the following time window and, in analogy, target ions of compounds eluting close to the beginning of a time window were monitored in the respective preceding time window (Table 3.5). This allowed quantification of the respective analytes even if the separation line of the time windows was not situated exactly between the peaks. For all the compounds the monitored ions were

Table 3.5 SIM conditions employed for quantitative LC/APCI/MS analysis of 23 benzodiazepins, flumazenil, zaleplone, zolpide and zopiclone

	Gain value	Fragmenter voltage (V)	Time windows	Analyte	Target ions (*m/z*)
Group 1			Window 1.1	Chlordiazepoxide	300
			(0–2.70 min)	Clonazepam	316
				Medazepam	271
				Midazolam	326
				Nordazepam	271
				Nordazepam-d_5	276
				Oxazepam	287
	3.0	100	Window 1.2	Clonazepam	316
			(2.71–5.00 min)	Desalkylflurazepam	289
				Midazolam	326
				Nordazepam	271
				Nordazepam-d_5	276
				Oxazepam	287
				Trimipramine-d_3	298
			Window 1.3	Camazepam	372
			(5.01–8.50 min)	Diazepam	285
				Diazepam-d_5	290
				Prazepam	325
				Trimipramine-d_3	298
Group 2	6.0	100	Window 2.1	Bromazepam	316
			(0–3.00 min)	Flumazenil	304
				Zolpidem	308
			Window 2.2	Temazepam	241
			(3.01–8.50 min)	Tetrazepam	289
Group 3	8.0	100	Window 3.1	Flurazepam	388
			(0–2.20 min)	Zaleplone	306
				Zopiclone	389
			Window 3.2	Alprazolam	309
			(2.21–3.70 min)	Flunitrazepam	314
				Flunitrazepam-d_7	321
				Flurazepam	388
				Lormetazepam	335
				Nitrazepam	282
				Triazolam	343
				Zaleplone	306
			Window 3.3	Flunitrazepam	314
			(3.71–8.50 min)	Flunitrazepam-d_7	321
				Lormetazepam	335
Group 4	4.0	200	Window 4.1	Lorazepam	303
			(2.00–3.15 min)		
			Window 4.2	Brotizolam	393
			(3.15–8.50 min)	Clobazam	259
				Metaclazepam	393

$[M+H]^+$ ionic species, except for brotizolam, clobazam, lorazepam and metaclazepam due to interferences with other analytes. Different labelled ISs have been employed because the use of trimipramine-d_3, a typical standard for liquid–liquid extraction of these compounds,[14] did not lead to sufficient validation data for all the analytes. The different IS were assigned to the different analytes in the following combination:

- diazepam-d_5 for zolpidem, midazolam, lormetazolam, tetrazepam and camazepam;
- flunitrazepam-d_7 for zopiclone, bromazepam, flurazepam, desalkylflurazepam, oxazepam, lorazepam, nitrazepam and flunitrazepam;
- nordazepam-d_5 for flumazepam, chlordiazepoxide, zaleplone, medazepam, clonazepam, nordazepam, temazepam, brotizolam, metaclazepam, diazepam and prazepam;
- trimipramine-d_3 for alprazolam, triazolam and clobazam.

The analytical procedure was validated according to the internationally accepted recommendation and interfering peaks due to common drugs typically taken in combination with analytes could be excluded basing on their different r.t. values. This is clearly shown in Figure 3.14 where typical MID chromatograms of an extract of plasma sample, spiked with IS and the analytes, is reported. The response for all the analytes was linear from subtherapeutic to overdose concentration (Table 3.6). A weighted ($1/c^2$, where c is the concentration) least-squares model was used for calculation of calibration curves to account for the unequal variances (heteroscedasticity) across the calibration range. The low- and high-level recoveries ranged from 87.4 to 113.6 % for all the analytes but bromazepam. It should be emphasized that all LODs determined in full-scan mode were at least equal to the corresponding LOQs in MID mode. The LOQs corresponded to the lowest concentrations used for the calibration curves with a signal-to-noise ratio of at least 10.

This multi-drug assay has proven to be applicable to more than 100 authentic plasma samples in daily routine work. As an example, the MID chromatogram obtained from an extract of an authentic plasma sample, analysed for compliance monitoring, is reported in Figure 3.15. As can be seen, well resolved peaks due to oxazepram, lorazepam, nitrazepam and clonazepam were detected: the amount of these analytes was estimated to be 0.08, 0.01, 0.04 and 0.03 mg/L, respectively.

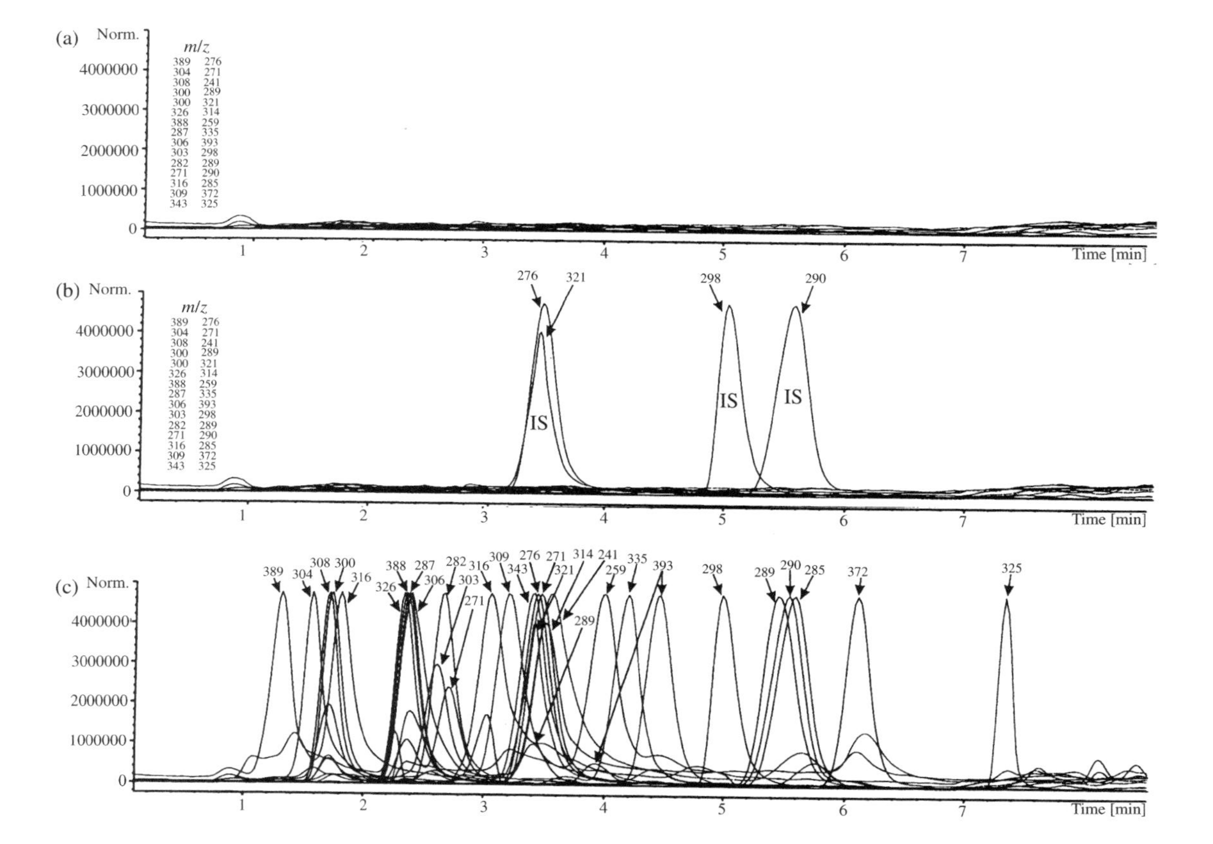

Figure 3.14 Smoothed and merged SIM chromatograms normalized with respect to the most abundant IS trimipramine-d_3 of a blank plasma extract (a), of a blank plasma extract spiked with 0.2 mg/L diazepam-d_5, 0.01 mg/L flunitrazepam-d_7, 0.2 mg/L nordiazepam-d_5 and 1.0 mg/L trimipramine-d_3 (b) and of a blank plasma spiked with ISs and the analytes at therapeutic concentrations (c). Reproduced from Kratzsch *et al.* (2004)[11] with permission from John Wiley & Sons, Ltd

Table 3.6 Therapeutic concentration range, linearity, LOD, LOQ and recovery data of the LC/APCI/MS assay for the studied drugs

Drug	Therapeutic concentration range (mg/L)	Linearity range (mg/L)	Coefficient of determination (R^2)	Limits (mg/L)		Recovery	
				LOD (scan mode)	LOQ (SIM mode)	Concentration (mg/L)	Mean ± SD (%)
Alprazolam	0.005–0.12	0.0025–0.15	0.998	0.0025	0.0025	0.005	96.0 ± 2.6
						0.12	96.0 ± 6.2
Bromazepam	0.08–0.2	0.04–0.25	0.551	0.01	0.04	0.08	44.6 ± 55.1
						0.2	39.4 ± 81.3
Brotizolam	0.001–0.03	0.0005–0.0375	0.989	0.0005	0.0005	0.001	104.1 ± 4.4
						0.03	99.4 ± 5.8
Camazepam	0.1–0.6	0.05–0.75	0.990	0.01	0.05	0.1	104.5 ± 2.7
						0.6	104.1 ± 6.0
Chlordiazepoxide	0.4–3.0	0.2–3.75	0.996	0.01	0.2	0.4	104.5 ± 3.1
						3.0	102.6 ± 1.5
Clobazam	0.1–1.2	0.05–1.5	0.969	0.01	0.05	0.1	87.4 ± 3.5
						1.2	97.6 ± 6.2
Clonazepam	0.01–0.05	0.005–0.0065	0.982	0.005	0.005	0.01	106.7 ± 2.1
						0.05	97.9 ± 3.4
Diazepam	0.2–2.0	0.1–2.5	0.992	0.01	0.1	0.2	100.5 ± 7.3
						2.0	98.7 ± 1.2
Flumazenil	0.02–0.1	0.01–0.125	0.991	0.01	0.01	0.02	102.5 ± 2.1
						0.1	94.9 ± 3.9
Flunitrazepam	0.01–0.03	0.005–0.0375	0.991	0.005	0.005	0.01	104.1 ± 2.3
						0.03	99.4 ± 4.4
Flurazepam	0.005–0.15	0.0025–0.1875	0.993	0.0025	0.0025	0.005	102.1 ± 7.5
						0.15	91.1 ± 5.7
Desalkylflurazepam	0.005–0.15	0.0025–0.1875	0.992	0.005	0.0025	0.005	103.9 ± 7.0
						0.15	98.6 ± 5.5
Lorazepam	0.02–0.25	0.01–0.3125	0.997	0.02	0.01	0.02	101.3 ± 4.2
						0.25	101.7 ± 6.7

Lormetazepam	0.001–0.03	0.0005–0.0375	0.987	0.0005	0.0005	0.001	101.9 ± 5.9
						0.03	97.9 ± 4.7
Medazepam	0.001–0.03	0.0005–0.0375	0.987	0.0005	0.0005	0.001	101.9 ± 5.9
						0.03	97.9 ± 4.7
Metaclazepam	0.05–0.2	0.025–0.25	0.992	0.025	0.025	0.05	103.0 ± 2.2
						0.2	92.1 ± 4.2
Midazolam	0.04–0.25	0.02–0.5	0.996	0.005	0.02	0.04	101.2 ± 2.3
						0.4	99.5 ± 5.3
Nitrazepam	0.004–0.2	0.002–0.25	0.994	0.002	0.002	0.004	98.5 ± 6.8
						0.2	97.9 ± 4.6
Nordazepam	0.2–2.0	0.1–2.5	0.991	<0.01	0.1	0.2	103.4 ± 2.1
						2.0	100.7 ± 4.3
Oxazepam	0.02–2.0	0.01–2.5	0.992	0.001	0.01	0.02	104.7 ± 1.1
						2.0	99.6 ± 5.6
Prazepam	0.01–2.0	0.005–2.5	0.995	0.001	0.005	0.01	93.5 ± 6.2
						2.0	97.6 ± 5.1
Temazepam	0.02–1.0	0.01–1.25	0.991	0.005	0.01	0.02	104.9 ± 1.9
						1.0	89.1 ± 8.1
Tetrazepam	0.05–0.6	0.025–0.75	0.993	0.005	0.025	0.05	105.1 ± 2.3
						0.6	100.7 ± 3.6
Triazolam	0.002–0.1	0.001–0.125	0.997	0.001	0.001	0.002	96.0 ± 6.1
						0.1	96.1 ± 6.7
Zaleplone	0.01–0.1	0.005–0.125	0.992	0.005	0.005	0.01	97.9 ± 3.8
						0.1	99.7 ± 3.7
Zolpidem	0.08–0.4	0.04–0.5	0.990	0.01	0.04	0.08	103.3 ± 2.4
						0.4	104.1 ± 3.3
Zopiclone	0.01–0.15	0.005–0.1875	0.997	0.0025	0.005	0.01	113.6 ± 3.5
						0.15	109.8 ± 11.4

LOD, limit of detection; LOQ, limit of quantitation.

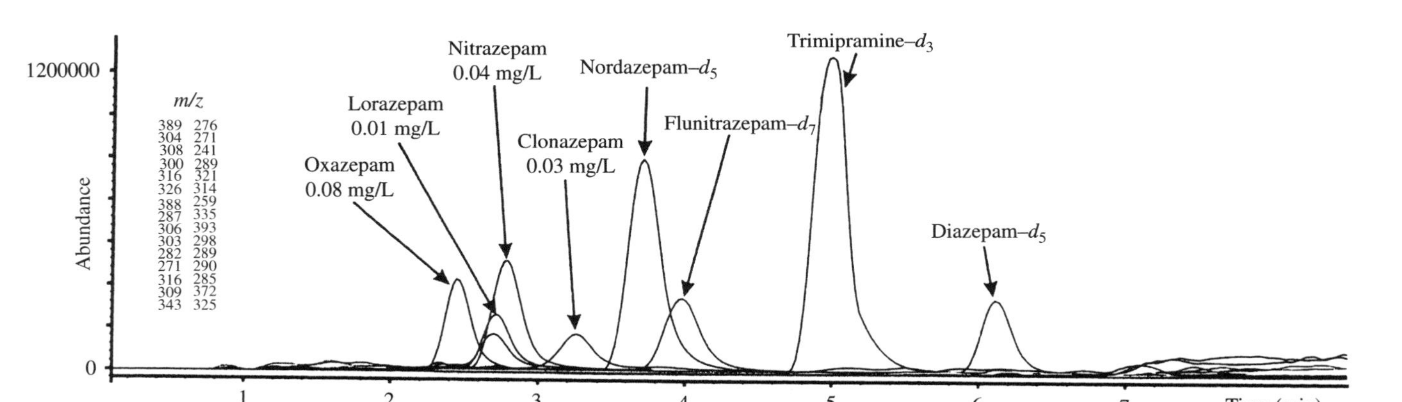

Figure 3.15 Smoothed and merged chromatograms of an extract of an authentic plasma sample submitted for compliance monitoring indicating 0.03 mg/L of clonazepam, 0.01 mg/L of lorazepam, 0.04 mg/L of nitazepam and 0.08 mg/L of oxazepam. Reproduced from Kratzsch *et al.* (2004)[11] with permission from John Wiley & Sons, Ltd

3.2 CHOICE OF A SUITABLE IONIZATION METHOD

As described in Chapter 1, different ionization methods are nowadays available and their choice is mainly related to the physico-chemical properties of the analytes of interest, as well as to the chromatographic system employed for the analyses.

In the case of a complex matrix, it reasonable to assume that it is composed of many different classes of compounds, exhibiting physico-chemical and thermodynamic properties different from each other.

These properties can be validly employed to develop a method exhibiting high selectivity. As discussed in Section 1.1.2, gas-phase protonation reactions:

$$M + AH^+ \rightarrow MH^+ + A$$

take place only if the proton affinity of M is higher than that of A. Consequently, the choice of a suitable reactant AH^+ can make possible the selective ionization of M, leaving neutral all the molecular species with proton affinity values lower than that of A. In other words, it is possible to use gas-phase protonation by CI experiments to detect only analytes with PA values higher than or equal to that of the reactant A.

This approach becomes more selective in the case of negative ion formation through electron attachment or deprotonation reaction:

$$M + e^- \rightarrow M^{-\cdot}$$
$$M + B^- \rightarrow [M - H]^- + BH$$

As an example of the complementarity of the results which can be achieved by the use of different ionization methods, the positive and negative ion spectra of a secondary metabolite of *Aspergillus Wentii* Wehmer[15] (namely 15-acetoxyhexadecylcitraconic anhydride, compound **1**, Figure 3.16), are reported in Figure 3.17.

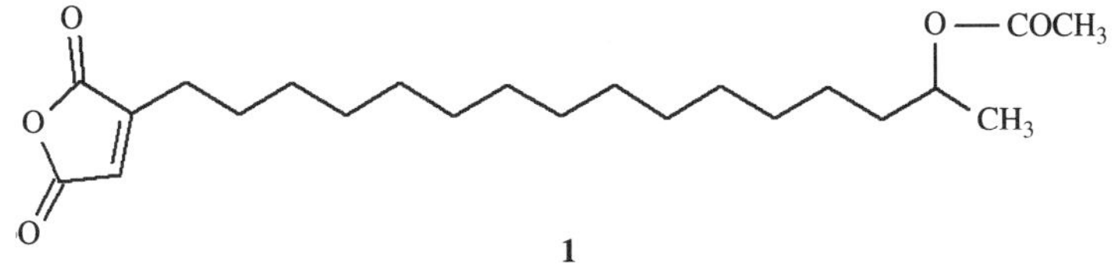

O—$COCH_3$

CH_3

1

Figure 3.16 Structure of 15-acetoxyhexadecylcitraconic anhydride (**1**)

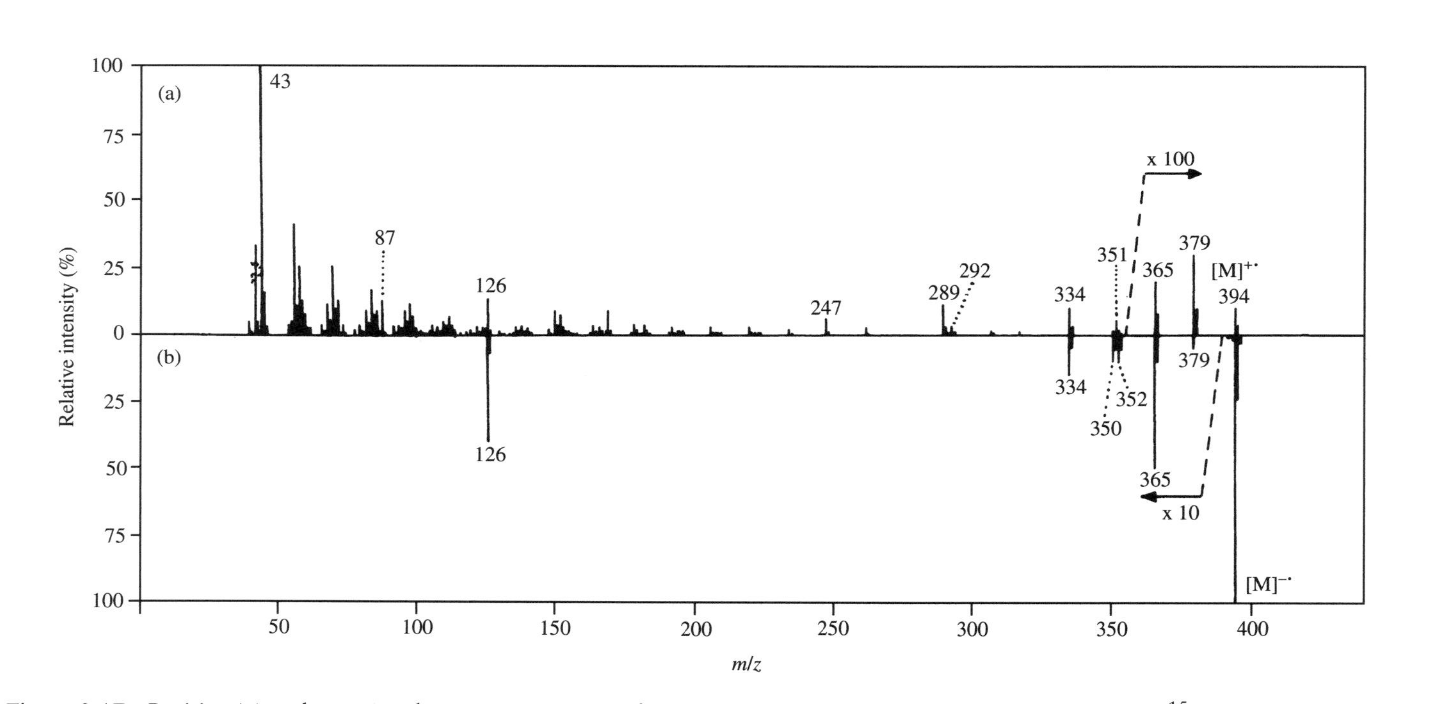

Figure 3.17 Positive (a) and negative (b) ion mass spectra of compound **1**. Reproduced from Selva *et al.* (1980)[15] with permission from John Wiley & Sons, Ltd

As can be seen, the positive ion mass spectrum shows diagnostic ions of very low abundance (the $M^{+\cdot}$ species exhibits a relative abundance of 0.01 %), while in negative ion mode (obtained by electron attachment reactions) the $M^{-\cdot}$ species represent the base peak of the spectrum and only a few fragments (all well related to the molecular structure) are detectable. This behaviour is in agreement with the high stability of the molecular anions of maleic anhydride derivatives, which can be formed by thermal electron and secondary electron capture, leading to long-lived species by nuclear excited resonance processes, and is also in agreement with the charge localization concept.

This approach can be employed also to obtain a fingerprinting of molecules of interest present in complex matrices without the need for chromatographic separation. The data obtained in the identification of dicarboxylic aciduria, an inborn error of metabolism, are a good example of this approach.[16] Dicarboxylic aciduria may arise from several inherited or acquired conditions which cause an enzymatic blockage or a metabolitic overflow of mitochondrial β-oxidation of medium chain monocarboxylic fatty acid. The biochemical consequence is a shunt towards a microsomal o-oxidation and synthesis of the corresponding dicarboxylic acids, particularly sebacic (C10:0), suberic (C8:0) and adipic (C6:0) acids. Figure 3.18 shows the GC/MS profile of the urinary organic acids of a 3-year-old boy affected by progressive neuromuscular deterioration.

The power of NICI(OH^-) for a rapid and direct mapping of urinary organic acids is shown by the data reported in Figure 3.19. Figure 3.19(a) shows the electron impact spectrum of the whole acidic fraction of the sample investigated. Most of the TIC is due to low mass ionic

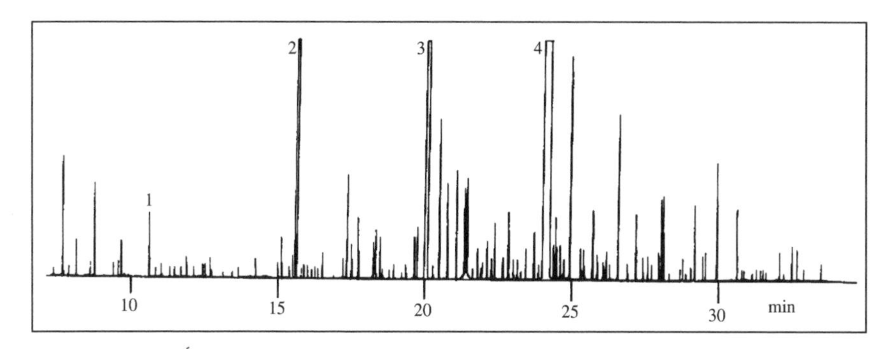

Figure 3.18 TIC chromatogram of the urinary organic acid extract of a 3-year-old boy affected by progressive neuromuscular deterioration. Peak 1, succinic acid; peak 2, adipic acid; peak 3, suberic acid; peak 4, sebacic acid. Reproduced from Rinaldo *et al.* (1985)[16] with permission from John Wiley & Sons, Ltd

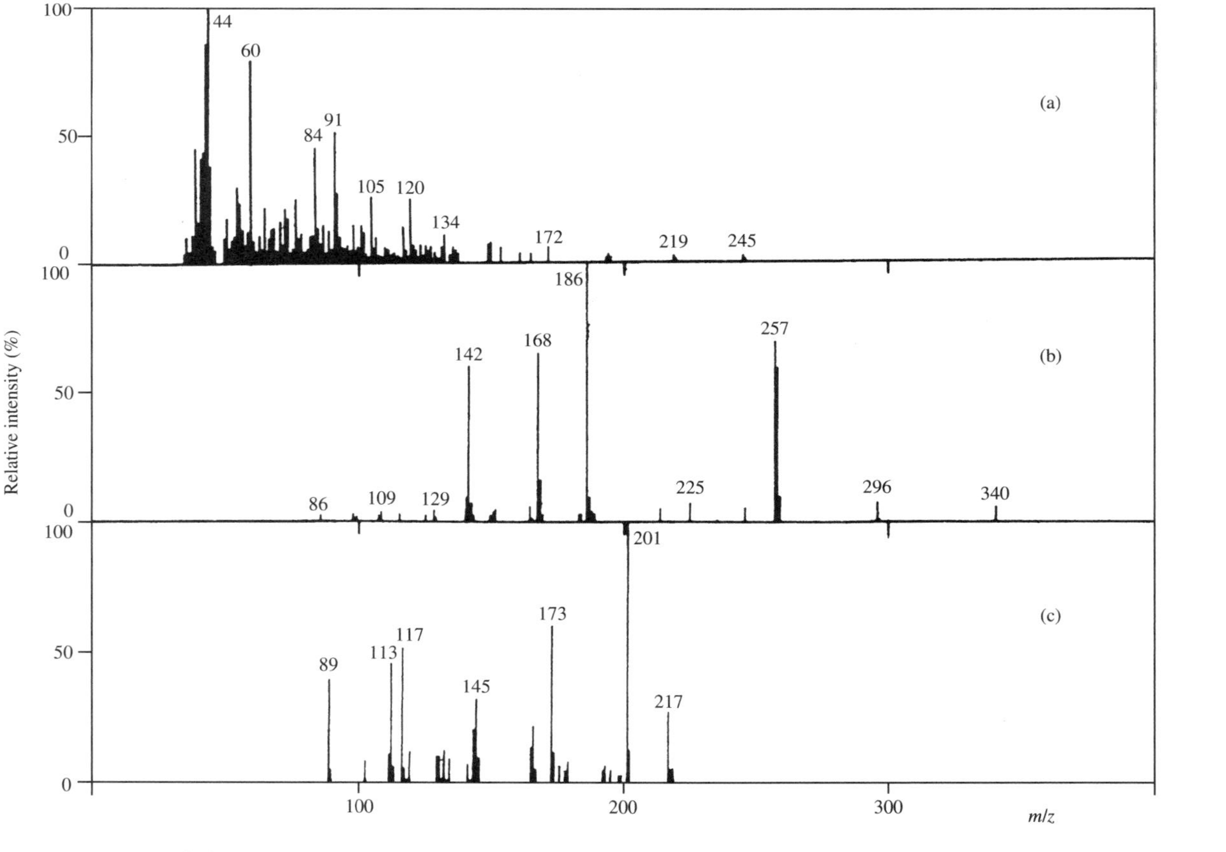

Figure 3.19 Mass spectra of the nonderivatized urinary extract leading to the chromatogram of Figure 3.18 obtained under: (a) EI conditions; (b) electron attachment conditions; (c) NICI(OH^-) conditions. Reproduced from Rinaldo *et al.* (1985)[16] with permission from John Wiley & Sons, Ltd

species, while no molecular peaks of the dicarboxylic acids are shown. Figure 3.19(b) illustrates the negative ion mass spectrum of the same urinary fraction, obtained under electron attachment conditions. Molecular species of the dicarboxylic moieties are still not detectable, though ionic species at higher mass value are present. This result can be explained by the fact that electron attachment underlines molecular species with high electron affinity. The abundant ions present in the spectrum (Figure 3.19b) do not correspond necessarily to the molecular species present at highest rates in the mixture, but only to moieties exhibiting the highest electron affinity.

A qualitative profile of the urinary acidic fraction can be clearly obtained only under hydroxyl negative ion chemical ionization [NICI(OH^-)] conditions. Figure 3.19(c) illustrates the forecast $[M - H]^-$ species of dicarboxylic acids detected by GC/MS, which are easily detected (sebacic acid at *m/z* 201, suberic acid at *m/z* 173, adipic acid at *m/z* 145 and succinic acid at *m/z* 117). In addition, other ions are present at *m/z* 89, presumably due to lactic acid.

Summarizing the data shown in Figure 3.19, it follows that:

(i) EI does not lead to any significant results due either to the copresence of a large number of compounds or to the ineffectiveness of the method in detecting carboxylic acid molecular ions.
(ii) Electron attachment provides evidence of the compounds exhibiting the highest electron affinity present in urine, but not the acids of interest.
(iii) NICI performed by reaction with gas-phase generated OH^-, strongly favours the classic acid–base reaction:

$$RCOOH + OH^- \rightarrow RCOO^- + H^2O$$

with the formation of acid anions.

What is observed in analytical methods requiring sample vaporization (EI and CI) can be transferred to more recently developed ionization methods operating with sample solution, in particular in electrospray conditions.

As an example, the results obtained in the analysis by ESI of a raw extract of artichoke of phytochemical interest are reported in Figure 3.20. The electrospray spectrum obtained in positive ion mode is highly complex, with peaks practically at every *m/z* value. Furthermore the most abundant peaks do not correspond to any compound described among the main components of the natural extract. In contrast, the

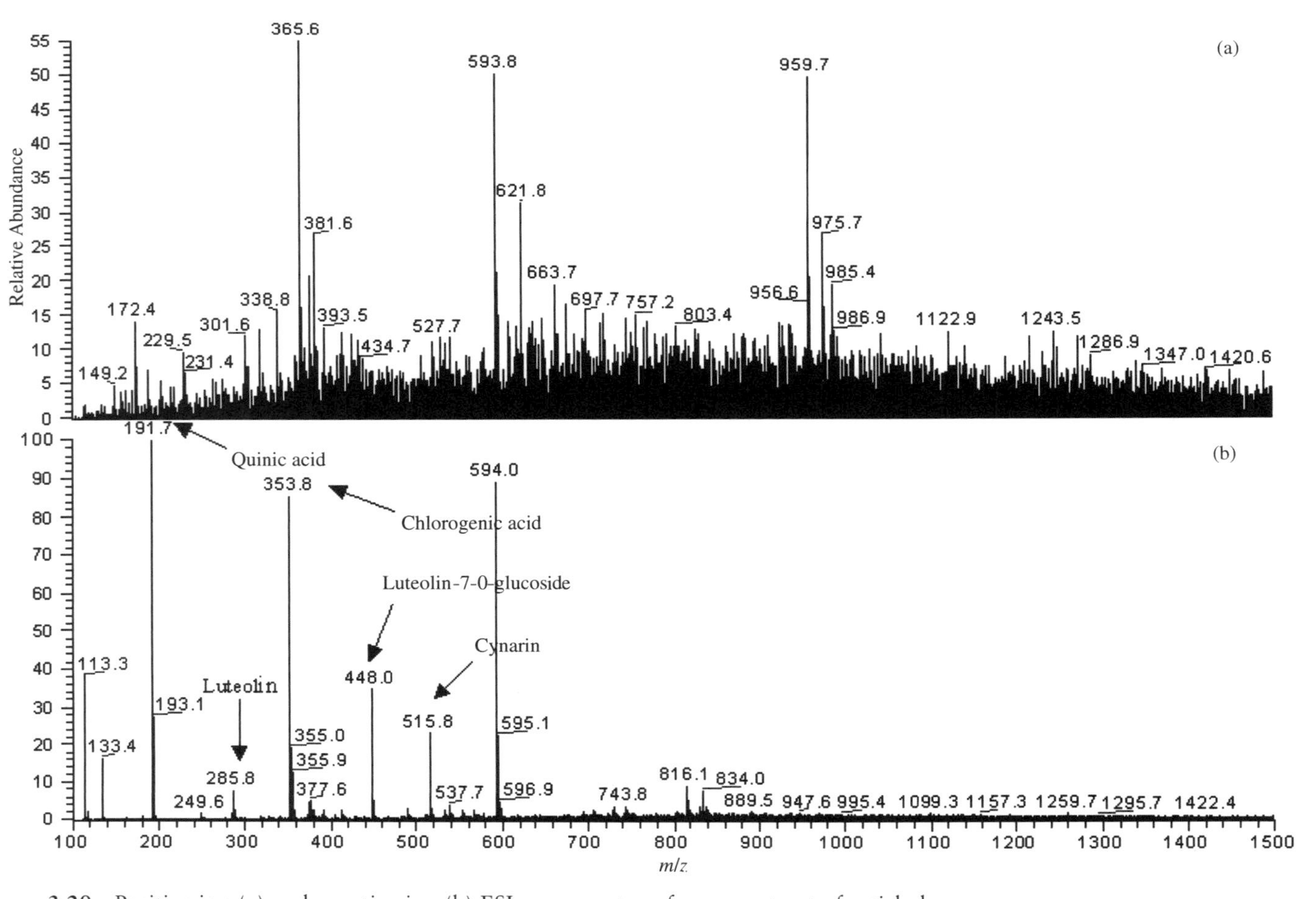

Figure 3.20 Positive ion (a) and negative ion (b) ESI mass spectra of a raw extract of artichoke

negative ion spectrum shows only a few, well defined peaks due to molecular anions typical of this botanic species.

In all the examples reported above a major point must be emphasized: in negative ion spectra, apart from the identification of structurally diagnostic ions, a wide decrease in so-called 'chemical noise' (i.e. the chemical background due to the presence of many ionic species) is observed, with a consequent increase in the signal-to noise ratio.

3.3 AN EXAMPLE OF HIGH SPECIFICITY AND SELECTIVITY METHODS: THE DIOXIN ANALYSIS

3.3.1 Use of High Resolution MID Analysis

A clear example of the improvement in sensitivity and selectivity, which can be obtained by the use of MID in high resolution conditions in quantitative analysis, is the analytical method employed for the determination of polychlorodibenzodioxin (PCDD) and polychlorodibenzofuran (PCDF) congeners (PCDD/F) in different matrices. The structures and the list of the toxicologically more relevant PCDD/F are reported in Figure 3.21.

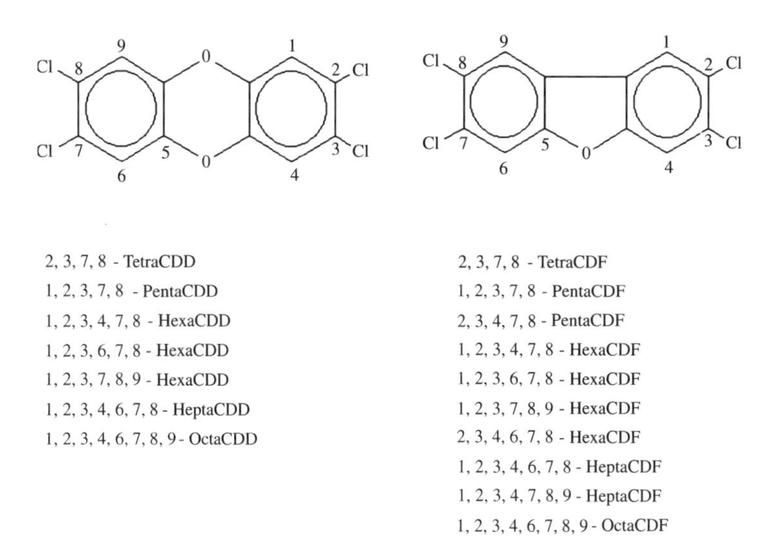

Figure 3.21 Structures of the toxicologically more relevant tetra, penta, hexa, hepta and octa CDD and CDF

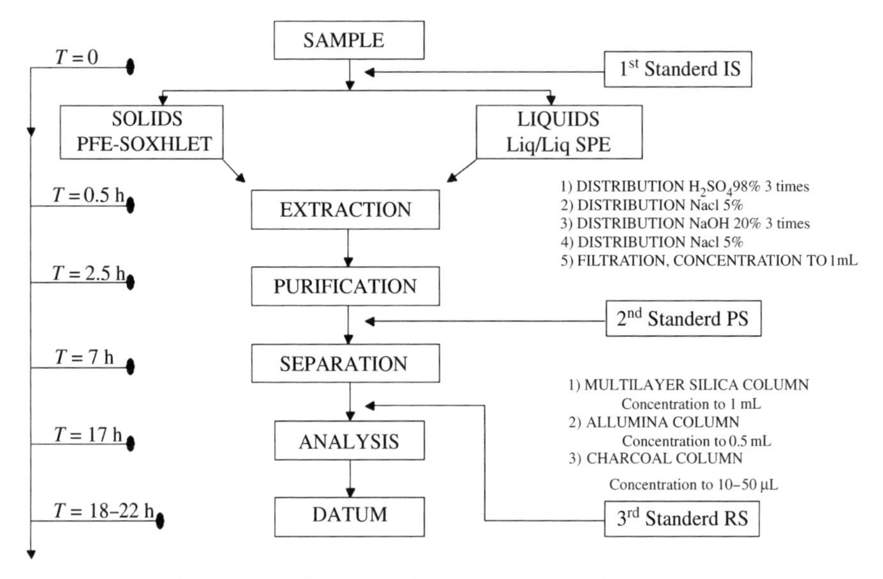

Figure 3.22 Flow chart of the sample treatment usually employed for PCDD/F analysis

It should be emphasized that this approach fully satisfies the US EPA 1613 (Revision B), the European standard EN-1948-1 and the Ontario Ministry of Environment (MOE) methods, all requiring the monitoring of the total concentration of all PCDD/F by high resolution gas chromatography/high resolution mass spectrometry (HRGC/HRMS) at ppq levels.

The first problem in this kind of analysis arises from the analytical sample preparation. The flow chart related to sample treatment is reported in Figure 3.22. Two extraction procedures can be employed, depending on the nature of the substrate of interest (solid or liquid) and the related results can give an evaluation of extraction procedure. Extraction–purification–separation phases are required and three different labelled standards are introduced in different steps of the purification treatment:

- IS (15 congeners of $2,3,7,8\text{-}^{13}C_{12}$ from tetraCDD/F to octaCDD, isotopic purity $\geq 99\,\%$);
- purification standard (PS) (2,3,7,8-tetra$^{13}C_{12}$DD, isotopic purity $\geq 96\,\%$);

- recovery standard (RS) (1,2,3,4-tetra$^{13}C_{12}$DD, 1,2,3,7,8,9-Hexa$^{13}C_{12}$DD, isotopic purity $\geq$99 %).

From their levels, it is possible to evaluate the extraction yield.

The time scale, reported on the left of Figure 3.22, should be noted: the whole procedure requires aproximately 22 h!

The example reported below concerns the analysis of a certified sediment, with a total content of PCDD and PCDF of 20042 ± 4706 pg/g. The instrument used was a double focusing mass spectrometer, operating in EI positive ion mode. The instrumental parameters were accurately optimized, in order to obtain a mass resolution of 12 000 and the ions used for the different MID windows are reported in Table 3.7. The electron beam energy employed was 45 eV (instead of the commonly employed 70 eV).

The TIC chromatogram obtained by full scan analysis is reported in Figure 3.23. As can be seen, the baseline is very high, suggesting that many compounds are present in the extract and are not separated by the developed chromatographic procedure. The RIC obtained for the M^+ of tetraCDD (at *m*/*z* 320; accurate mass value 319.8965) is reported in Figure 3.24. Taking into account that under the GC conditions used the 2,3,7,8-tetraCDD r.t. is 30–31 min, it is impossible to distinguish the tetrachlorodioxin peaks from the background. The same results are obtained for all PCDD/F.

When the GC/MS analysis is carried out under high resolution MID conditions (Table 3.7), the RIC reported in Figure 3.25 is obtained. A large increase in signal-to-noise ratio is obtained. In the RIC chromatogram based on MID data related to $M^{+\cdot}$ ions of tetraCDD, an abundant and well resolved chromatographic peak is obtained at r.t. 30.52 min, as shown in Figure 3.26(a). This peak corresponds to 2,3,7,8-tetraCDD. The RIC chromatogram related to $M^{+\cdot}$ ions of 2,3,7,8-tetra$^{13}C_{12}$DD (used as IS) and of 1,2,3,4-tetra$^{13}C_{12}$DD (used as RS) are reported in Figure 3.26(b). It should be emphasized that in each MID analysis an accurate mass range is applied in order to determine the ions due to both the analyte and to the corresponding labelled ISs. Once the 2,3,7,8-tetraCDD/IS area ratio is determined, the calibration curve allows the analyte concentration in the original sample to be obtained and its recovery during the purification procedure to be determined by measuring the IS/RS peak ratio. Analogous results have been obtained for all PCDD/F congeners.

Table 3.7 MID time windows employed in high resolution GC/high resolution MS of PCDD/F

No.	Start (min)	End (min)	Accurate mass	Ion identification	Compound
			303.9016	M	TetraCDF
			305.8987	M+2	TetraCDF
			315.9419	M	Tetra$^{13}C_{12}$DF (IS or RS)
			317.9389	M+2	Tetra$^{13}C_{12}$DF (IS or RS)
1	28.30	34.00	319.8965	M	TetraCDD
			321.8936	M+2	TetraCDD
			331.9368	M	Tetra$^{13}C_{12}$DD (IS or RS)
			333.9338	M+2	Tetra$^{13}C_{12}$DD (IS or RS)
			354.9792[a]	LOCK[a]	PFK[a]
			339.8598	M+2	PentaCDF
			341.8569	M+4	PentaCDF
			351.9000	M+2	Penta$^{13}C_{12}$DF (IS or RS)
			353.8970	M+4	Penta$^{13}C_{12}$DF (IS or RS)
2	34.00	38.00	355.8547	M+2	PentaCDD
			357.8518	M+4	PentaCDD
			367.8949	M+2	Penta$^{13}C_{12}$DD (IS or RS)
			369.8919	M+4	Penta$^{13}C_{12}$DD (IS or RS)
			354.9792[a]	LOCK[a]	PFK[a]
			373.8208	M+2	HexaCDF
			375.8178	M+4	HexaCDF
			383.8639	M	Hexa$^{13}C_{12}$DF (IS or RS)
			385.8609	M+2	Hexa$^{13}C_{12}$DF (IS or RS)
3	38.00	41.30	389.8156	M+2	HexaCDD
			391.8127	M+4	HexaCDD
			401.8559	M+2	Hexa$^{13}C_{12}$DD (IS or RS)
			403.8529	M+4	Hexa$^{13}C_{12}$DD (IS or RS)
			430.9728[a]	LOCK[a]	PFK[a]
			407.7818	M+2	HeptaCDF
			409.7788	M+4	HeptaCDF
			417.8250	M	Hepta$^{13}C_{12}$DF (IS or RS)
			419.8220	M+2	HeptaCDF
4	41.30	48.00	423.7767	M+2	HeptaCDD
			425.7738	M+4	HeptaCDD
			435.8169	M+2	Hepta$^{13}C_{12}$DD (IS or RS)
			437.814	M+4	Hepta$^{13}C_{12}$DD (IS or RS)
			430.9728[a]	LOCK[a]	PFK[a]
			441.7428	M+2	OctaCDF
			443.7399	M+4	OctaCDF
			457.7377	M+2	OctaCDD
5	48.00	55.00	459.7348	M+4	OctaCDD
			469.7780	M+2	Octa$^{13}C_{12}$DD (IS or RS)
			471.7750	M+4	Octa$^{13}C_{12}$DD (IS or RS)
			442.9728[a]	LOCK[a]	PFK[a]

[a]PFK, perfluorokerosene used as mass calibrant.

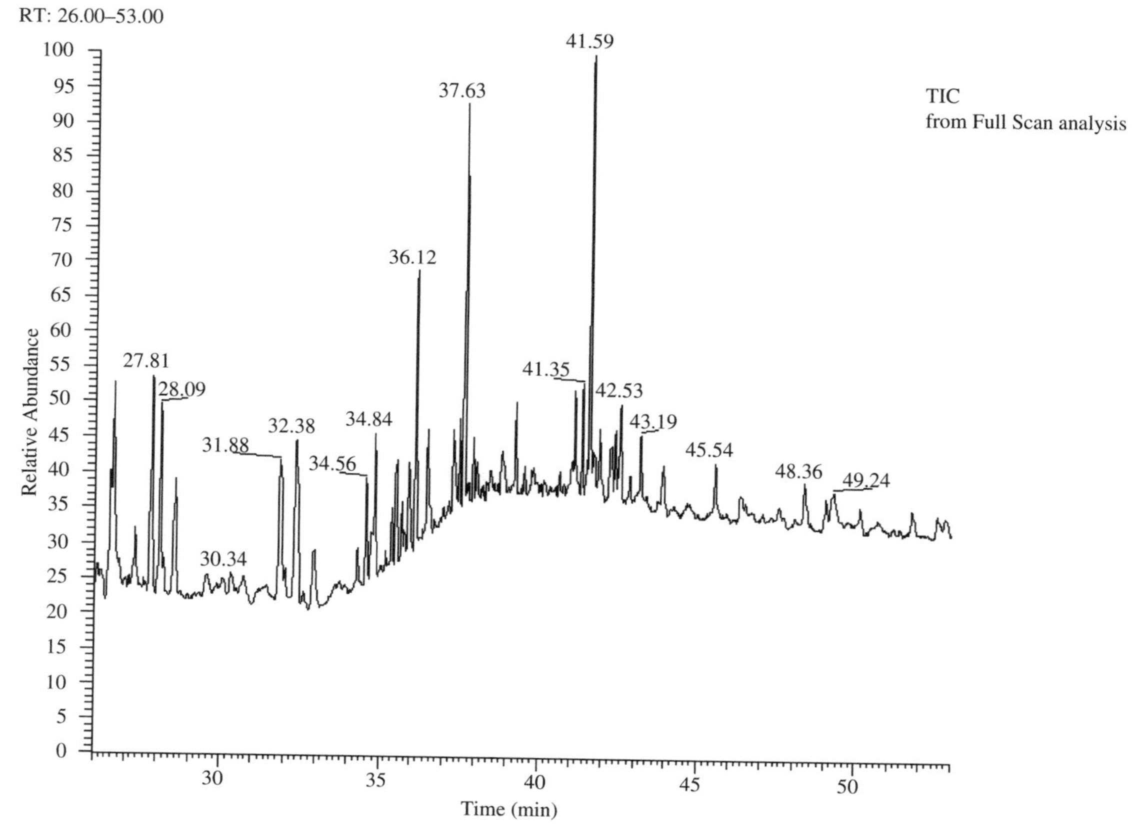

Figure 3:23 TIC chromatogram of a certified sediment with a total content of PCDD/F of 20.042 ± 4.706 pg/g

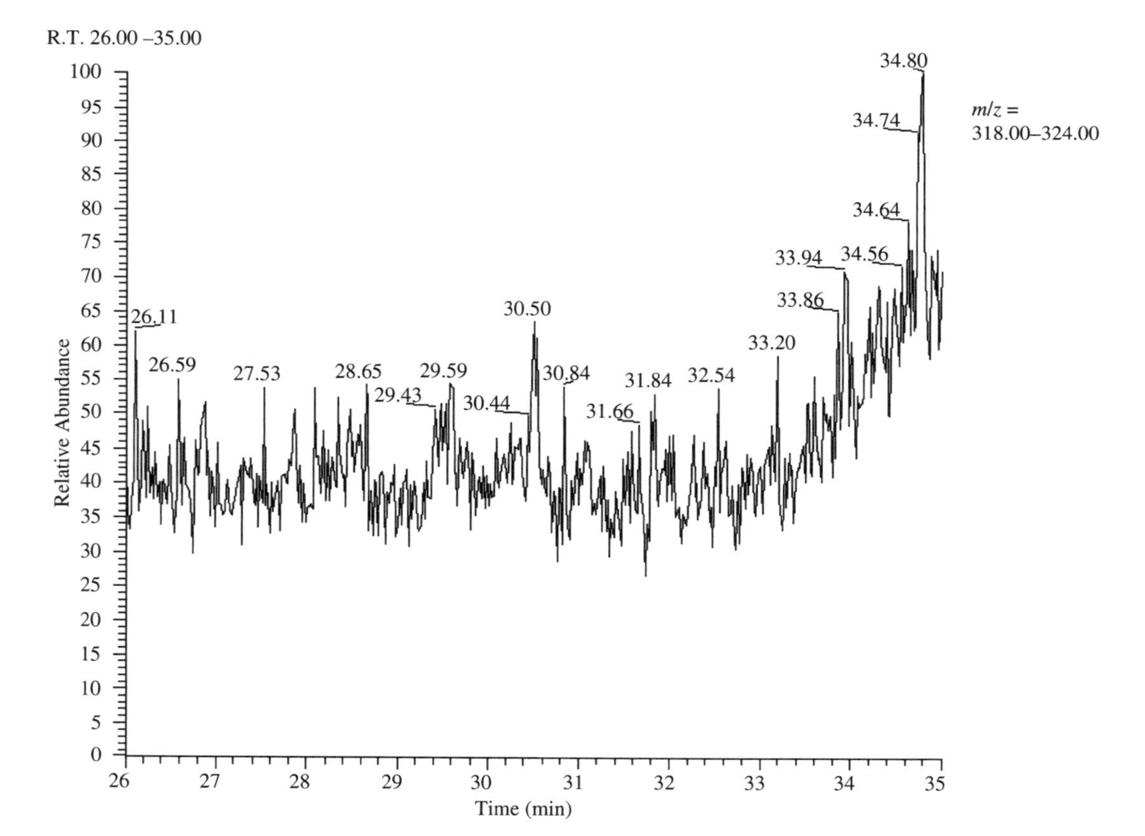

Figure 3.24 RIC related to $M^{+\cdot}$ of tetraCDD (*m/z* 320) obtained by elaboration of the data leading to the TIC chromatogram of Figure 3.23

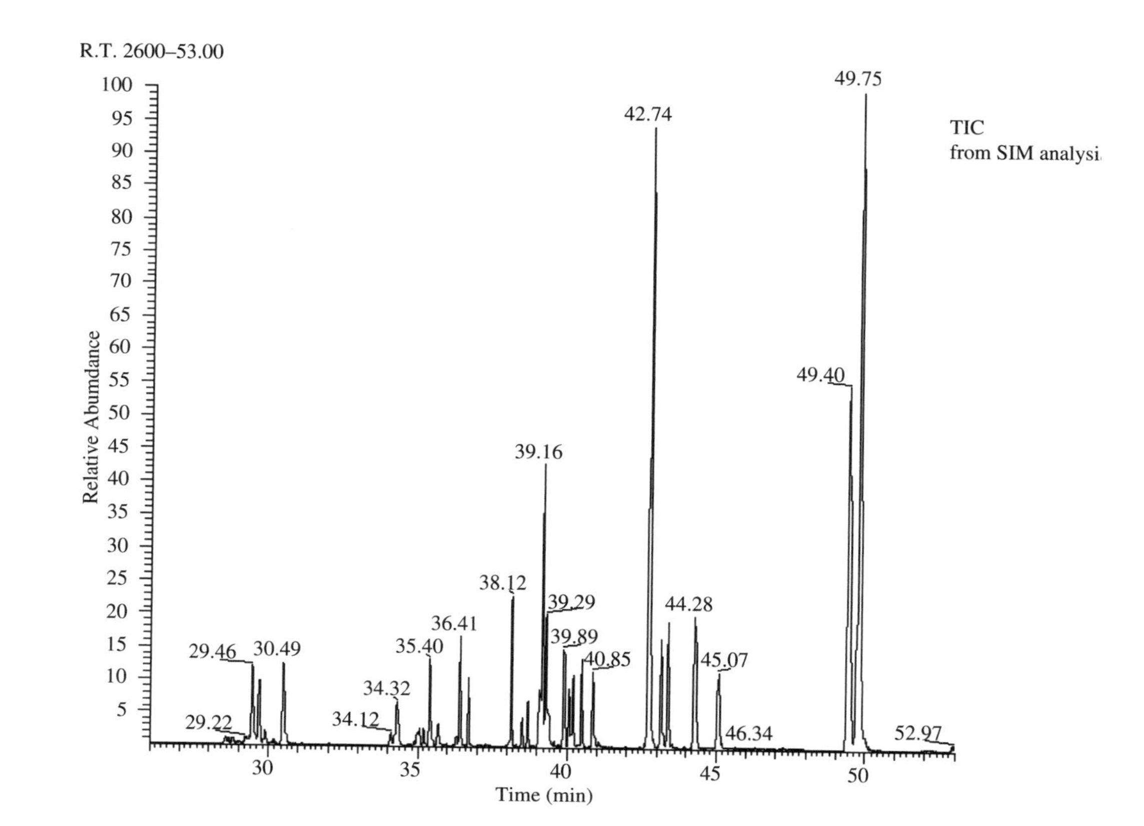

Figure 3.25 MID chromatogram of the certified sediment extract obtained in high resolution conditions

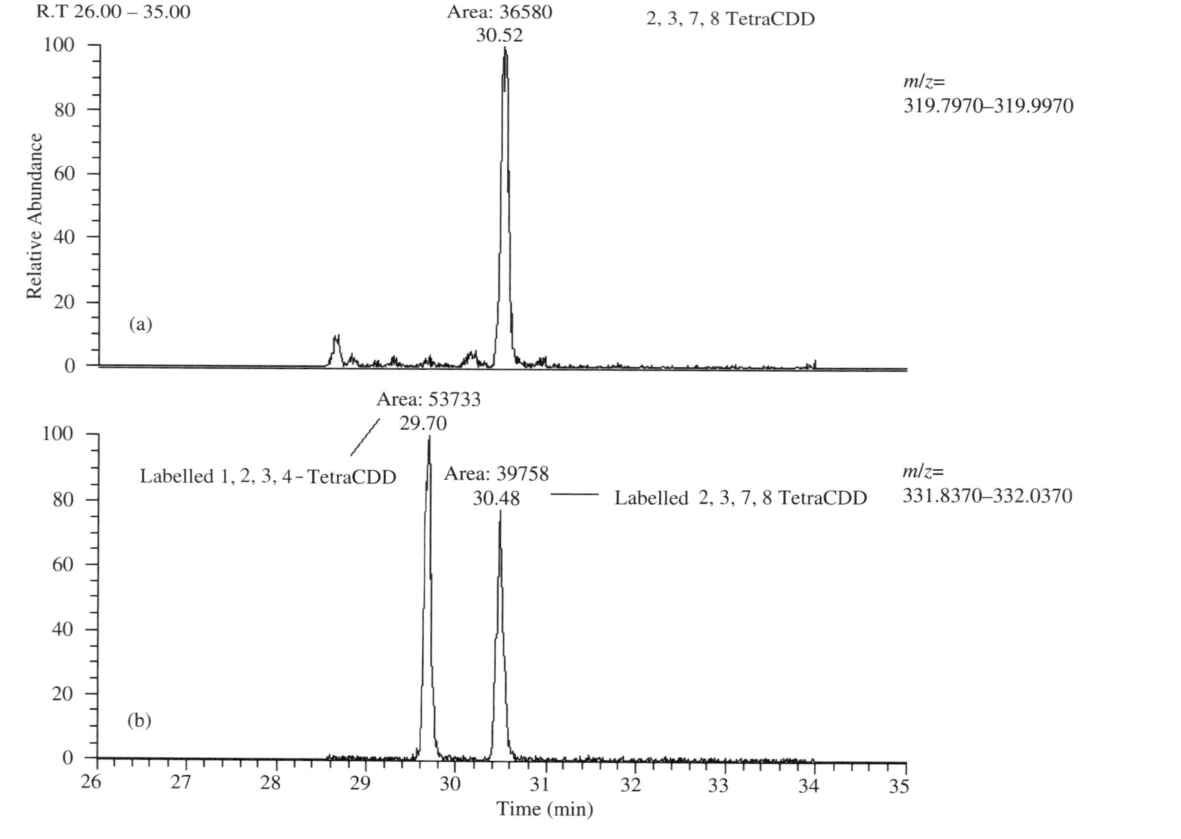

Figure 3.26 RIC chromatogram, based on the high resolution MID data, related to $M^{+\cdot}$ ions of tetraCDD (a); the peak detected at 30.52 min corresponds to 2,3,7,8-tetraCDD. RIC chromatogram related to $M^{+\cdot}$ ions of 2,3,7,8-$T^{13}C_{12}DD$ (used as IS) and of 1,2,3,4-$T^{13}C_{12}DD$ (used as RS) (b)

3.3.2 NICI in the Analysis of Dioxins, Furans and PCBs[17]

As discussed in Chapter 1, the specificity of a MS method can be increased by employing selective ionization methods or the MS/MS approach. These have also been employed in the case of polychlorinated dibenzo-*p*-dioxins, dibenzofurans and biphenyls.

The power of NICI has been tested by the use of different reagent gases. In the case of methane, thermal electrons are generated by EI of neutral methane, mainly through two different mechanisms:

$$CH_4 + e_p \rightarrow CH_4^{+\cdot} + e_t + e_p$$

$$CH_4 + e_p \rightarrow CH_3^{+} + H^{\cdot} + e_t + e_p$$

where e_p represents a primary electron and e_t is the thermal electron released from the methane molecule. The thermal electron interacts with the neutral molecule M of the analyte, giving rise to an odd electron molecular anion $M^{-\cdot}$:

$$M + e_t \rightarrow M^{-\cdot}.$$

The yield of these processes is to some extent dependent on experimental conditions (temperature, ion source type, reactant gas pressure, traces of oxygen, electron energy, concentration of the analyte), which need to be carefully optimized. Oehme and colleagues[18] compared the data obtained by EI and NICI. Detection limits in the range 0.2–3 pg were found for 4,6-chlorinated PCDFs, octaCDF, octaCDD, pentaCDD and hexaCDD in NICI conditions, while higher values (3–20 pg) where found in EI conditions.

More recently a comparative analysis of dioxins and furans performed by high resolution, electron ionization mass spectrometry and NICI has been carried out by Koester *et al.*,[19] demonstrating the positive and negative aspects of the two approaches. Comparable detection limits were obtained (100–200 fg for EI, 50–100 fg for NICI). Other studies have been carried out using O_2, CH_4/O_2, CH_4/N_2O, $iC_4H_{10}/CH_2Cl_2O_2$, Ar/CH_4, H_2, Ar, Xe, SF_6, and He as reagent gas. The results obtained, summarized in Table 3.8, have been critically evaluated and compared by Chernetsova *et al.*[17]

Table 3.8 Detection limits and characteristic ions in NCI mass spectra of PCDDs, PCDFs and PCBs with different reagent gases

Reagent gases	PCDDs	PCDFs	PCBs
CH_4	M^-, Cl^-, $[M-Cl]^-$, $[M-2Cl]^-$, $[M-H]^-$, $[M-3Cl]^-$, $[M-4Cl]^-$, $[M-xCl+xH]^-$ Detection limit 10^{-11}–10^{-14} g (except 2,3,7,8-tetraCDD)	M^-, Cl^-, $[M-2Cl]^-$, $[M-xCl+xH]^-$ Detection limit 10^{-10}–10^{-13} g	M^-, Cl^-, $[M-Cl]^-$, $[M-2Cl]^-$, $[M-3Cl]^-$, $[M-4Cl]^-$, $[M-H]^-$, $[M-Cl+H]^-$ Detection limit 10^{-12}–10^{-14} g
CH_4/O_2	M^-, $[M-Cl+O]^-$, $[M+O_2-C_6H_2Cl_2O_2]^-$ (*m/z* 176) Detection limit 10^{-10}–10^{-11} g. Possible to distinguish isomers	M^-, $[M-Cl+O]^-$, Cl^-, $[M-Cl+H]^-$	M^-, $[M-Cl+O]^-$, $[M-Cl]^-$, $[M-H]^-$, Cl^-, $[M-2Cl]^-$, $[M-Cl+H]^-$, $[M-HCl+O_2]^-$
O_2	M^-, $[M-Cl+O]^-$ $[M+O_2-C_6H_2Cl_2O_2]^-$ (*m/z* 176) Detection limit 10^{-10}–10^{-12} g. Possible to distinguish isomers	M^-, $[M-Cl+O]^-$, Cl^-	—
O_2 (APCI)	M^-, $[M-Cl+O]^-$, $[M+O_2-C_6H_2Cl_2O_2]^-$ (*m/z* 176), $[M-H]^-$ Detection limit 10^{-10}–10^{-13} g. Possible to distinguish isomers	$[M-Cl+O]^-$, Cl^-	—
CH_4/N_2O	M^-, $[M+OH]^-$, $[M+OH-Cl]^-$, $[M+O-Cl]^-$, $[M-Cl]^-$ Detection limit ca. 10^{-11} g for a few compounds. Possible to distinguish isomers	—	—
$i-C_4H_{10}/$ $/CH_2Cl_2/O_2$	$[M-Cl+O]^-$ Detection limit ca. 10^{-10} g without preliminary GC separation	$[M-Cl+O]^-$ Detection limit ca. 10^{-10} g without preliminary GC separation	$[M-Cl+O]^-$ Detection limit ca. 10^{-10} g without preliminary GC separation

Table 3.8 (*Continued*)

Reagent gases	PCDD_S	PCDF_S	PCB_S
Ar/CH_4	M^-, Cl^-, $[M-Cl]^-$, $[M-2Cl]^-$, $[M+H-Cl]^-$, $[M-3Cl]^-$ Detection limit ca. 10^{-12}–10^{-14} g	—	—
H_2	M^-, $[M\text{-}Cl]^-$, Cl^-; $[M\text{-}Cl+O]^-$ (in presence of O_2) Detection limit ca. 10^{-12} g for a few compounds	—	—
He	M^-, Cl^-, $[M-Cl]^-$	—	—
Ar	M^-, Cl^-, $[M-Cl]^-$	—	—
Xe	M^-, Cl^-	—	—
SF_6	Cl^-	—	—
NH_3	—	—	M^-, Cl^-, $[M-Cl]^-$, $[M-2Cl]^-$, $[M-3Cl]^-$, $[M-4Cl]^-$, $[M-Cl+H]^-$ Detection limit 10^{-12}–10^{-14} g
Ar/O_2	—	—	M^-, $[M\text{-}Cl+O]^-$, $[M-HCl+O_2]^-$ When the number of chlorine atoms <4, the sensitivity is better than using CH_4/O_2
CH_4/H_2O	—	—	M^-, $[M-Cl+O]^-$, $[M-HCl+O_2]^-$ When the number of chlorine atoms >4, the sensitivity is better than using CH_4/O_2

3.3.3 MS/MS in the Determination of Dioxins, Furans and PCBs[20]

The selectivity by high resolution MS and NICI can be reached also by MS/MS experiments. In an early paper by March *et al.*, a rapid screening technique, based on the MS/MS approach by an ion trap, was

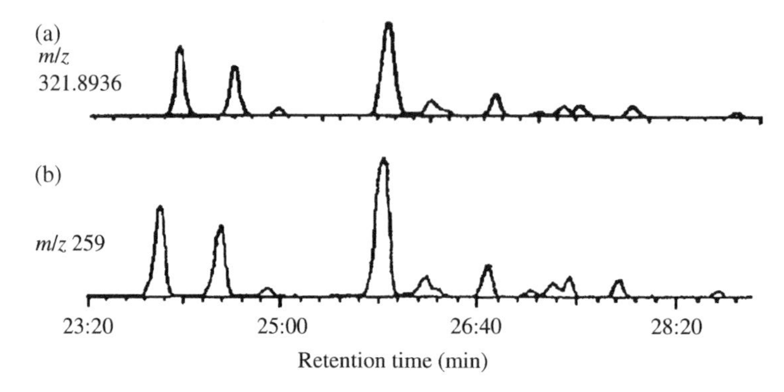

Figure 3.27 Two chromatograms for tetraCDDs obtained from a clam extract. (a) The chromatogram has been obtained in high resolution MID conditions for the $M^{+\cdot}$ ions at *m/z* 321.8936; (b) the chromatogram was obtained by monitoring the product ion signal (*m/z* 259) obtained by collision of the $M^{+\cdot}$ ions. Reproduced from March *et al.* (2000)[21] with permission from Elsevier

reported for the investigation and quantitative analysis of tetraCDD.[21] Figure 3.27 shows the chromatograms obtained by high resolution MS and by ion trap MS/MS experiments. The former originated by the ion signal obtained in SIM mode for *m/z* value 321.8936 ($[M+2]^{+\cdot}$ species of molecular ion), while the latter is due to the signal of the collisionally generated ion at *m/z* 259 (originating from a primary $COCl^{\cdot}$ loss). The two chromatograms are practically superimposable, indicating that the possible differences in critical energies related to the $COCl^{\cdot}$ loss from the different isomers does not affect the quantitative measurements.

The ion traps currently available allow multiple reaction monitoring (i.e. select different ions to be subjected to further collisions) to be performed and consequently product ions from tetra- to octaCDDs/CDFs can be monitored in a single chromatographic acquisition.

The scan function (i.e. the graphical representation of the variation of the potentials applied on the ion trap electrodes with respect to time) employed for MS/MS quantitative analysis of tetraCDD[22] is shown in Figure 3.28. The main RF voltage (applied on the ring electrode) is increased, to eject from the ion trap all ions with *m/z* values lower than 160. In order to avoid space-charge effects a first ionization stage (200 μs) is applied: it allows the optimum duration of the real ionization phase, which was always lower than 20 ms, to be determined. At this stage, all ions are trapped inside the device and by the application of a

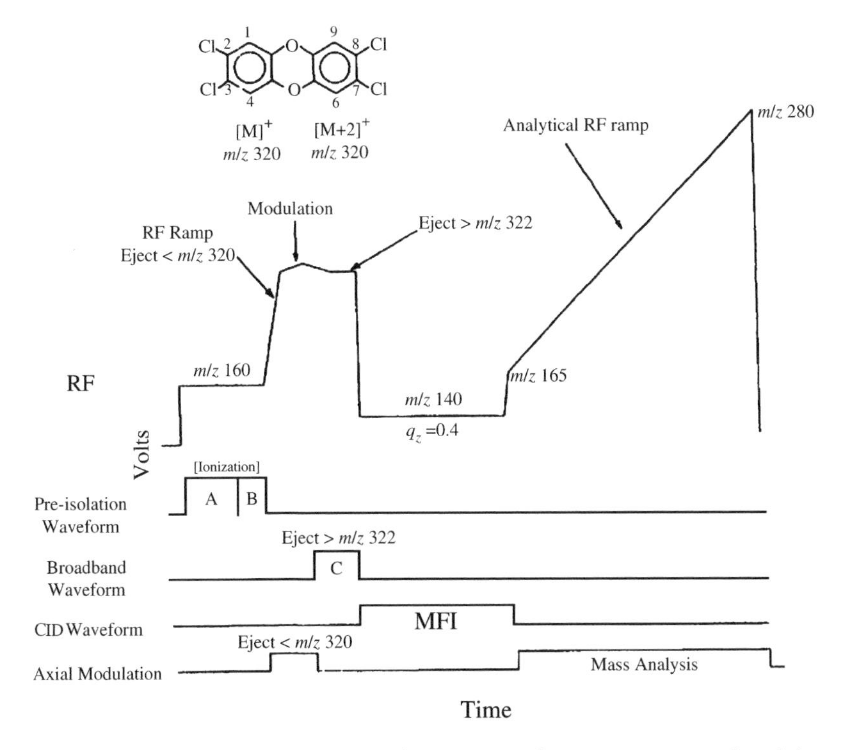

Figure 3.28 Scan function employed for MS/MS of tetraCDD. Reproduced from Splendore *et al.* (1997)[22] with permission from Elsevier

RF ramp (on the ring electrode) and of a broadband waveform (on the two end cap electrodes) all the ions with *m/z* values lower than 320 and higher than 322 are ejected from the trap. In these conditions $M^{+\cdot}$ and $[M+2]^{+\cdot}$ ions of tetraCDDs are the only ions present inside the trap. The application of a supplementary RF voltage on the two end caps with resonance frequency corresponding to the ions at *m/z* 320 and 322 leads to the activation of collisional phenomena of these ions with the He atoms present in the trap as buffer gas. The collisionally induced decomposition products are then ejected from the ion trap by increasing the main RF voltage and detected.

The scan function can be rapidly changed during the chromatographic run, so that different ions can be selected, decomposed and the related decomposition product ions detected.

The TIC of the mixture described in Table 3.9, is reported in Figure 3.29.[22] (The diamonds on the abscissa indicate the change of scan function.) The peak number–compound correspondence is

Table 3.9 Compilation of PCDDs and PCDFs and their relative concentration in solution S5

Compounds	Peak number	S5[a]
2,3,7,8-TetraCDD	3	200
2,3,7,8-TetraCDF	1	200
1,2,3,7,8-PentaCDD	6	1000
1,2,3,7,8-PentaCDF	4	1000
2,3,4,7,8-PentaCDF	5	1000
1,2,3,4,7,8-HexaCDD	10	1000
1,2,3,6,7,8-HexaCDD	11	1000
1,2,3,7,8,9-HexaCDD	12	1000
1,2,3,4,7,8-HexaCDF	7	1000
1,2,3,6,7,8-HexaCDF	8	1000
2,3,4,6,7,8-HexaCDF	9	1000
1,2,3,7,8,9-HexaCDF	13	1000
1,2,3,4,6,7,8-HeptaCDD	15	1000
1,2,3,4,6,7,8-HeptaCDF	14	1000
1,2,3,4,7,8,9-HeptaCDF	16	1000
OctaCDD	17	2000
OctaCDF	18	2000
$^{37}Cl_4$-2,3,7,8-TetraCDD	3	200
$^{13}C_{12}$-2,3,7,8-TetraCDD	3	100
$^{13}C_{12}$-1,2,3,4-TetraCDD	2	100
$^{13}C_{12}$-2,3,7,8-TetraCDD	1	100
$^{13}C_{12}$-1,2,3,7,8-PentaCDD	6	100
$^{13}C_{12}$-1,2,3,7,8-PentaCDF	4	100
$^{13}C_{12}$-2,3,4,7,8-PentaCDF	5	100
$^{13}C_{12}$-1,2,3,4,7,8-HexaCDD	10	100
$^{13}C_{12}$-1,2,3,6,7,8-HexaCDD	11	100
$^{13}C_{12}$-1,2,3,7,8,9-HexaCDD	12	100
$^{13}C_{12}$-1,2,3,4,7,8-HexaCDF	7	100
$^{13}C_{12}$-1,2,3,6,7,8-HexaCDF	8	100
$^{13}C_{12}$-2,3,4,6,7,8-HexaCDF	9	100
$^{13}C_{12}$-1,2,3,7,8,9-HexaCDF	13	100
$^{13}C_{12}$-1,2,3,4,6,7,8-HeptaCDD	15	100
$^{13}C_{12}$-1,2,3,4,6,7,8-HeptaCDF	14	100
$^{13}C_{12}$-1,2,3,4,7,8,9-HeptaCDF	16	100

[a]Concentrations (pg/μL) for native compounds are approximate only.

given in Table 3.9. As can be seen, the different compounds coelute with the related isotopically labelled standards. In this case, due to the high complexity of the molecular cluster originating from the distribution of natural isotopes of chlorine, the ISs are shifted by 12 Da with respect to the analyte, due to the presence of 12 ^{13}C atoms in the molecule.

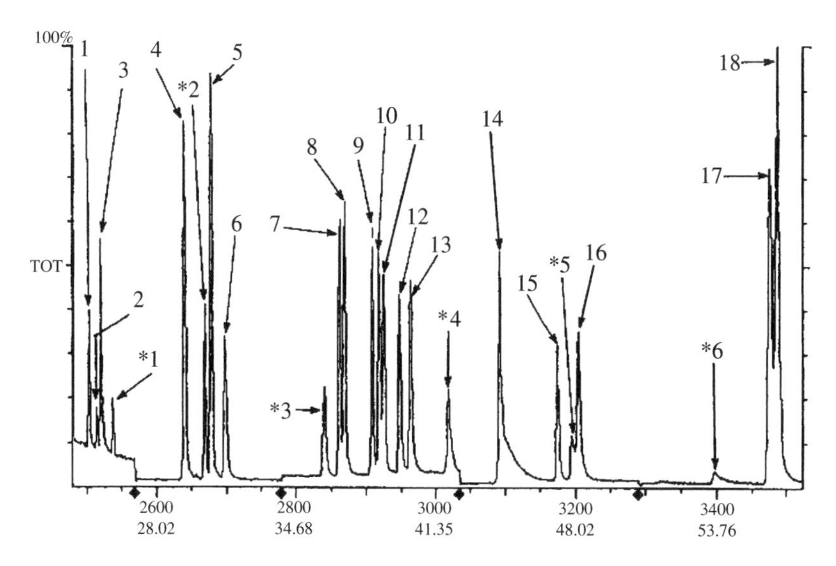

Figure 3.29 TIC chromatogram obtained after the mass spectrum merging procedure for PCDD/F contained in 1 μL of solution S5. The peak identification is given in Table 3.9. Reproduced from Splendore *et al.* (1997)[22] with permission from Elsevier

Thus, peak 1 is due to 2,3,7,8-tetraCDF and its coeluting ^{13}C-labelled isotopomer and consequently two different scan functions must be activated.

Typical data obtained by this approach are show in Figure 3.30, in which the marked tailing of the octachlorodioxin (peak 17 of Figure 3.29) and furan (peak 18 of Figure 3.29) are reported.[22] The selected ion chromatograms are related to the isolated molecular ion signal intensities (before any collision) in Figure 3.30(b) and to the fragment ion signal intensities obtained by collision in Figure 3.30(a) (see the related MS/MS spectra in the Figure).

An extensive comparison of the results obtained for the determination of PCDD/F by high resolution GC/high resolution MS, MS/MS experiments by both QQQ and MS/MS performed by ion trap has been reported in the literature.[20] In general, the high resolution GC/high resolution MS detection limit was lower than those of QQQ and ion trap, as shown in Figure 3.31. However, evidence was found that all interferences are not eliminated in high resolution GC/high resolution MS conditions and consequently MS/MS is of high interest.

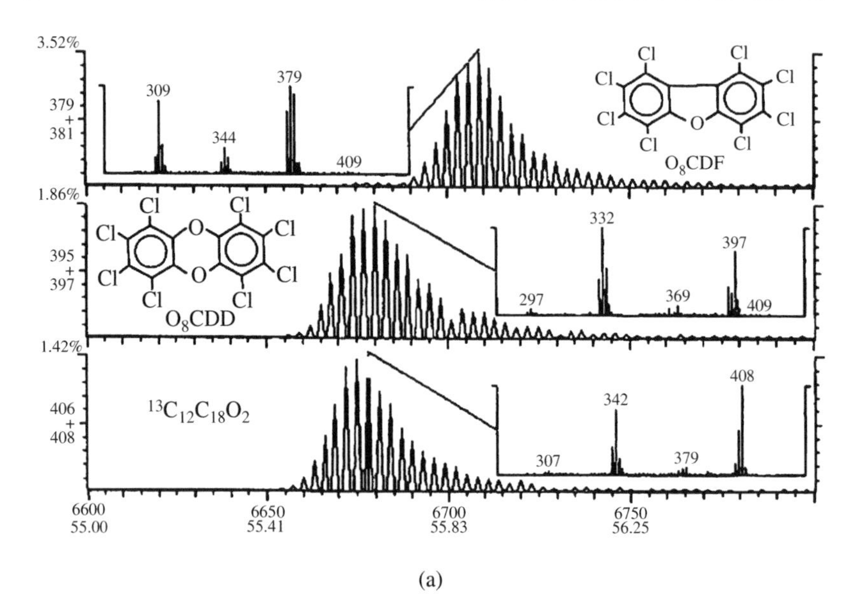

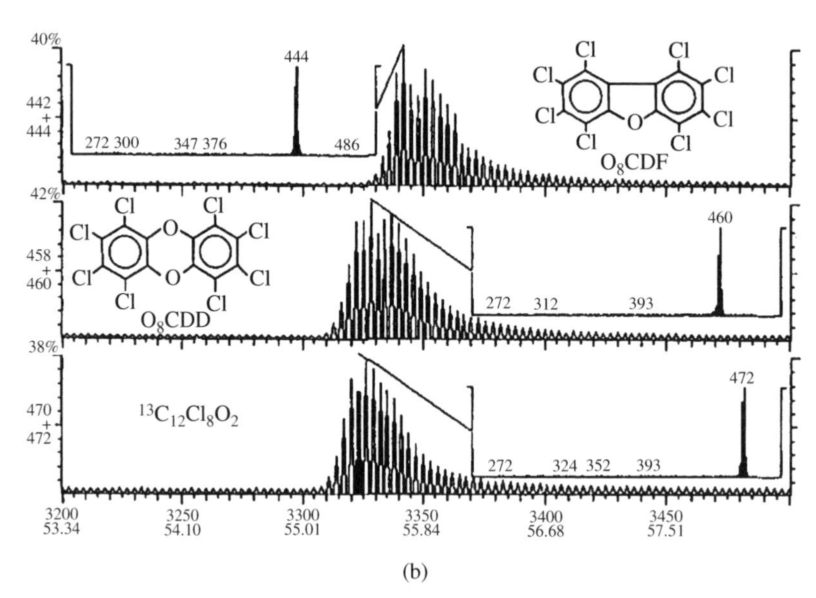

Figure 3.30 Unmerged ion chromatograms for three co-eluting octachlorinated compounds. Each chromatogram is composed of the signal intensities of the ionic species identified on the ordinate. (a) The fragment ion signal intensities were obtained by collisional spectra, and one such mass spectrum is displayed for each compound. (b) The isolated molecular ion signal intensities were obtained prior to collision, and a mass spectrum that shows the isolated molecular ions is displayed for each compound. Reproduced from Splendore *et al.* (1997)[22] with permission from Elsevier

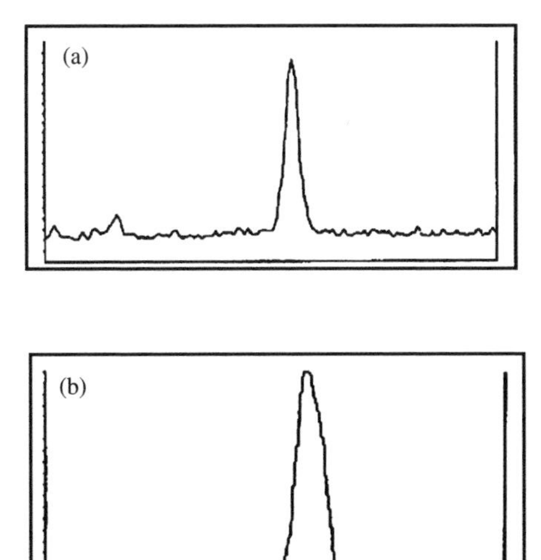

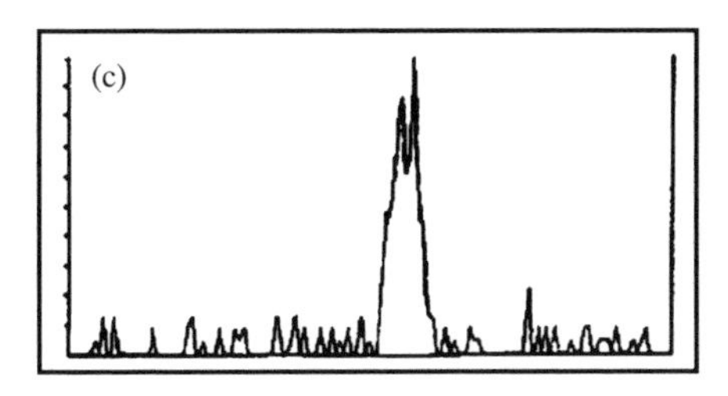

Figure 3.31 Ion signals obtained with each instrument for low concentrations of 2,3,7,8-tetraCDD: (a) high resolution conditions, 0.5 pg injected, the signal intensity sum due to *m/z* 320 and 322 is shown; (b) triple quadrupole, 1.0 pg injected, the signal intensity sum due to *m/z* 257 and 259 is shown; (c) ion trap, 0.5 pg injected, the signal intensity sum due to *m/z* 257, 259, 194 and 196 is shown. The signal-to-noise ratios are approximately 30 in the first case, 50 in the second and 15 in the last one. Reproduced from March *et al.* (2000)[21] with permission from Elsevier

3.4 AN EXAMPLE OF MALDI/MS IN QUANTITATIVE ANALYSIS OF POLYPEPTIDES: SUBSTANCE P

As described in Chapter 1, MALDI favours the formation of molecular species ($[M+H]^+$, $[M\text{-}H]^-$) and is particularly effective for the MS analysis of macromolecules, either from synthesis (polymers) or from natural substrates (proteins, oligonucleotides, polysaccharides).

As emphasized in Chapter 1, the mechanisms through which molecular species are generated are quite complex, based on both physical and chemical phenomena strongly reflected in the spectrum quality. For these reasons the use of MALDI for quantitative analysis seemed to be, in principle, practically impossible. However, recently some papers have appeared on MALDI application for quantitative analysis of biologically relevant polypeptides in a complex matrix, proving that, once accurate set-up of the different instrumental parameters has been performed, valid quantitative data can be obtained.

As an example of this approach, we report here the data obtained in the analysis of Substance P (SP) in rat brain.[23]

SP (Figure 3.32) is a peptide neurotransmitter distributed in the peripheral and central nervous system, where it acts as a neuromodulator. Abnormal levels of SP are associated with disorders such as schizophrenia and Huntington's, Parkinson's and Alzheimer's diseases. Its presence in a variety of tissues and its participation in various diseases make the quantitative analysis of SP extremely relevant.

MALDI/TOF MS has become a valuable and important analytical tool in biological research owing to its ability to analyse low concentrations of peptides and proteins in complex biological mixtures and it has been employed in the development of a quick method for detecting and determining the SP level in rat brain tissue using α-cyano-4-hydroxycinnamic acid (HCCA) as matrix. The method is based on analysing the SP level in the rat brain by immunoprecipitation, in which the SP polyclonal antibodies are raised, followed by solid phase extraction. The levels of SP were determined by measuring the abundance of $[M+H]^+$ ions by using the calibration curves constructed by external standards. The influence of the laser on the signal intensities and sample preparation techniques including flight time deviation for accurate mass measurements were evaluated. In combination with MALDI/MS, CZE analyses were also performed and the results obtained by the two analytical approaches were compared.

Due to the complex biological fluid mixtures (brain tissue extracts), it was necessary to isolate low concentrations of the analyte of interest from the biological fluid prior to the MALDI/MS analysis. Immunoaffinity capture of the analyte from the biological mixture is a method that pre-concentrates the analyte and provides the most selective form of analyte isolation.[24] High affinity polyclonal SP antibodies were used to specifically purify and enrich the SP. During incubation the SP was captured on the antibody, and after a series of washes the immunoprecipitated samples were subjected to a MALDI sample preparation

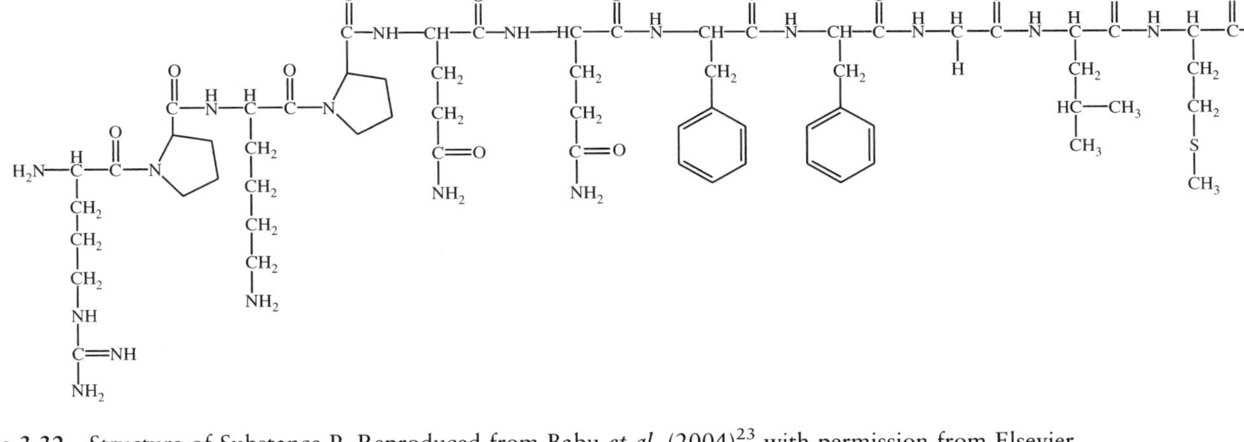

Figure 3.32 Structure of Substance P. Reproduced from Babu *et al.* (2004)[23] with permission from Elsevier

step. MS analysis of the immunoprecipitated sample confirmed the presence of SP for the detection of the related $[M+H]^+$ species at *m/z* 1348.8.

The laser power was found to be an important parameter, strongly affecting the peak $[M+H]^+$ intensity. MS analyses were performed with different laser power, ranging from the minimum intensity necessary for signal production to the maximum intensity, which was very close to the detector saturation. Through a series of experiments using SP concentration ranging from 50 to 250 fmol/μL, the laser intensity was set to the initial minimum energy and further increased until an acceptable and well reproducible signal was obtained.

In order to maximize the signal quality and reproducibility, the laser intensity, the sample preparation methods, the crystallization process and the *m/z* ranges were carefully adjusted. Ten samples were prepared for different HCCA/analyte combination. Every mass spectrum was recorded as the sum of 40 consecutive spectra collected from a selected spot on the probe tip and the average peak intensity values were taken for analysis.

Using this instrumental parameterization, the system was calibrated using angiotensin II (Ang II) and ACTH fragment 18-39 (ACTH) as external standards to normalize the SP $[M+H]^+$ peak intensities. For this, a working curve was a first generated for these external standards. Singly charged Ang II and ACTH $[M+H]^+$ ion peak intensities were used to generate the curve. Regression analysis resulted in the following linear equations:

$$y = 516.4x + 121.73 \text{ (Ang II)}$$
$$y = 560.17x + 79.32 \text{ (ACTH)}$$

with correlation coefficients of 0.9988 (Ang II) and 0.9899 (ACTH), respectively, where x is the Ang II and ACTH concentrations and y the analyte peak intensity.

The SP working curves were constructed for a standard SP concentration series by measuring the peak height mass spectral response as a function of SP concentration. Angiotensin II, ACTH (1 pmol/μL each) and different SP concentrations (from 0.5 to 20 pmol/μL) were premixed with the matrix. SP $[M+H]^+$ peak intensity for the different SP concentrations was normalized to the ($[\text{Ang II}+H]^+ + [\text{ACTH}+H]^+$) peaks, and the normalized response was plotted as a function of the SP concentration. The SP working curve response (Figure 3.33) followed a linear relationship ($y = 0.1144x + 0.2145$) over the concentration range

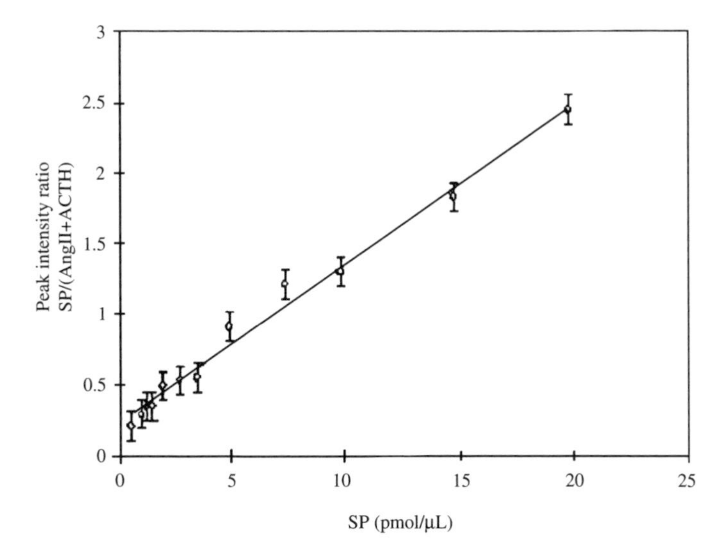

Figure 3.33 SP calibration curve. $[SP+H]^+$ion peak intensity (normalized to $[M+H]^+$ peak intensity of angiotensin II and ACTH) versus SP concentration. Reproduced from Babu *et al.* (2004)[23] with permission from Elsevier

used with a correlation coefficient of 0.9988 and a 0.74 % relative standard deviation.

The immunoprecipitated samples (A and B) from brain tissue were subjected to MS analysis, and, as mentioned earlier, the mass spectrum was acquired at fixed MALDI conditions. The singly charged $[M+H]^+$ ion peak corresponding to the SP molecular mass was observed. The SP concentrations were determined in the brain tissue extracts from two individuals and are reported in Table 3.10, together with the data obtained for the same sample by CZE (ultraviolet detection), showing good agreement between the results achieved by the two techniques.

Table 3.10 MALDI/MS and CZE quantification of SP in the rat brain

	Substance P (pmol/μg protein)	
Immunoprecipitated samples	**MALDI/MS** ($n = 10$)	**CZE** ($n = 3$)
A	0.12 ± 0.06	0.23 ± 0.04
B	0.36 ± 0.05	0.38 ± 0.02

Protein concentration (μg/μL): sample A = 2.495; sample B = 2.201.

REFERENCES

1. A. Braithwaite and F. J. Smith in *Chromatographic methods*, 4th Edition, Chapman and Hall, New York (1985).
2. L. R. Snyder and J. J. Kirkland in *Introduction to Modern Liquid Chromatography*, 2nd Edition, John Wiley & Sons, Inc., New York (1979).
3. M. Khaledi (Ed.) *High Performance Capillary Electrophoresis: Theory, Techniques and Applications*, John Wiley & Sons, Inc., New York (1998).
4. R. M. Smith (Ed.) *Supercritical Fluid Chromatography*, The Royal Society of Chemistry, London (1989).
5. A. Lapolla, R. Flamini, T. Tonus, D. Fedele, A. Senesi, R. Reitano, E. Marotta, G. Pace, R. Seraglia and P. Traldi, *Rapid Commun. Mass Spectrom.*, **17**, 876–878 (2003).
6. R. Flamini, G. De Luca and R. Di Stefano, *Vitis*, **41**, 107–114 (2002).
7. G. De Revel and A. Berthirand, *J. Sci. Food Agric.*, **61**, 267–272 (1993).
8. A. Lapolla, R. Flamini, A. Dalla Vedova, A. Senesi, R. Reitano, D. Fedele, E. Basso, R. Seraglia and P. Traldi, *Clin. Chem. Lab. Med.*, **41**, 1166–1173 (2003).
9. J. D. Henion and G. A. Maylin, *Biomed. Mass Spectrom.*, **7**, 115–121 (1980).
10. S. Dai, H. Song, G. Dou, X. Qian, Y. Zhang, Y. Cai, X. Liu and Z. Tang, *Rapid Commun. Mass Spectrom.*, **19**, 1273–1282 (2005).
11. C. Kratzsch, O. Tenberhen, F. T. Peters, A. A. Weber, T. Kraemer and H.H. Maurer, *J. Mass Spectrom.*, **39**, 856–872 (2004).
12. H. H. Maurer, C. J. Schmitt, A. A. Weber and T. Kraemer, *J. Chromatogr.B*, **748**, 125–135 (2000).
13. K. A. Mortier, K. M. Clauwaert, W. E. Lambert, J. F. Bocxlaer, E. G. Van den Eeckhout, C. H. Van Peteghen and A. P. De Leenheer, *Rapid Commun. Mass Spectrom.*, **15**, 1773–1775 (2001).
14. H. H. Maurer in *Mass Spectral and GC Data of Drugs, Poisons, Pesticides, Pollutants and their Metabolites, Part 4*, K. Pfleger, H. H. Maurer and A. Weber (Eds), John Wiley & Sons, Ltd, Weinheim (2000).
15. A. Selva, P. Traldi, L. Camarda and G. Nasini, *Biomed. Mass Spectrom.*, **7**, 148–152 (1980).
16. P. Rinaldo, G. Miolo, L. Chiandetti, F. Zacchello, S. Daolio and P. Traldi, *Biomed. Mass Spectrom.*, **12**, 570–576 (1985).
17. E. S. Chernetsova, A. I. Revelsky, I. A. Revelsky, I. A. Mikhasenko and T.G. Sobolevsky, *Mass Spectrom. Rev.*, **21**, 373–387 (2002).
18. M. Oehme, S. Manø, A. Mikalsen and P. Kirschmer, *Chemosphere*, **15**, 607–617 (1986).
19. C. J. Koester, R. L. Harless and R. H. Hites, *Chemosphere*, **24**, 421–426 (1992).
20. J. B. Plomley, H. Lausevic and R. E. March, *Mass Spectrom. Rev.*, **19**, 305–365 (2000).
21. R. E. March, M. Splendore, E. J. Reiner, R. S. Mercer, J. B. Plomley, D. S. Waddell and K. A. McPherson, *Int. J. Mass Spectrom. Ion Processes*, **197**, 283–297 (2000).
22. M. Splendore, J. B. Plomley, R. E. March and R. S. Mercer, *Int. J. Mass Spectrom. Ion Processes*, **165/166**, 595–609 (1997).
23. C. V. S. Babu, J. Lee, D. S. Lho and Y. S. Yoo, *J. Chromatogr. B*, **807**, 307–313 (2004).
24. R. W. Nelson, J. S. Krone, A. C. Bieber and P. Williams, *Anal. Chem.* **67**, 1153–1158 (1995).

4

Some Thoughts on Calibration and Data Analysis

Quantitative analysis and estimation of detection limits in the concentration domain require, in general, a previous calibration step. Therefore in this chapter we will describe and discuss thoroughly the calibration procedure and we will adapt it, following statistical arguments, to the problem of the determination of the critical level L_C, of the limit of detection L_D and of the limit of quantification L_Q.[1–7] We here will mainly refer to the papers of Gibbons and colleagues[8] and of Schwartz[9–10] for the definitions and the calculation procedure adopted.

As outlined in Chapter 2, the samples used in the calibration design in GC/MS analysis are solutions at various known concentrations of the analyte and at constant known concentration of the IS. Consequently, the dependent variable y in the calibration plot is the ratio of the analyte and of the IS responses, and the independent variable x is the known concentration ratio. This procedure is valid under the basic assumption that the error in response ratio is greater than that in the concentration ratio of the standard solutions. Otherwise, the reverse, a regression of concentration ratios on the response ratios, must be carried out.

We will treat both straight line and quadratic models since a curvature in the calibration plot can arise for two main reasons:[11] a large dynamic range and the eventual use of an IS of type A, isotopically labelled compound. It is well known that, in the case of a not isotopically pure standard, the curve of the isotope ratio (ratio of the ion current observed for the unlabelled species to that of the labelled species) versus the molar

Quantitative Applications of Mass Spectrometry I. Lavagnini, F. Magno, R. Seraglia and P. Traldi

ratio (ratio of the unlabelled to labelled compounds) is theoretically curvilinear.[12]

The same two reasons imply very often the non constant variance of the response ratios at different concentration ratios.[13] Consequently ordinary least-squares regression must be substituted for weighted least-squares procedure, where the inverse of the variance is used as weighting factor.

4.1 CALIBRATION DESIGNS

It is noteworthy to outline that the calibration design used for the estimation of detection limits can be quite different[1,2,5] from that used for a quantitative analysis. In the former case the known concentrations of the spiked solutions must be in the range of the hypothesized detection limit, whereas in the latter the standard solutions must bracket the unknown concentration. In particular, some authors suggest that the blank solution response must be inserted in the regression procedure when the detection limits are required. It is common opinion that the number of suitably spaced calibration points can range between seven and ten to obtain a reliable model for the calibration line[1,6,8] and that the basic pattern must be repeated several times, at least eight to ten replicates[7,14] to gain information about the overall variances of the responses at the various concentrations. This requirement is due to the fact that a considerable source of variability is always the preparation of the sample and of the calibration solutions.

4.2 HOMOSCEDASTIC AND HETEROSCEDASTIC DATA

Ordinary least squares (OLS) regression is used with data of uniform variance, the homoscedastic case, whereas weighted least squares (WLS) regression is appropriate under heteroscedastic conditions to draw the right model from the calibration data. It is known that WLS does not greatly affect the parameter estimates but it has a deep effect on the estimates of the precision and consequently on the detection limit estimate.[9] To verify whether homoscedasticity or heteroscedasticity exists two approaches can be followed: (i) Bartlett's test is done to compare the variances of the replicates at different concentration levels;[15] (ii) the plot of residuals fitting an unweighted least squares line versus predicted values is examined. A funnel shape indicates an increasing variability with increasing response magnitude, i.e. with concentration. A more immediate analysis of the uniformity of the experimental variances with concentration is simply the plot of the differences of each replicate from the mean of the replicates at each concentration. Values close to zero indicate uniformity in the variance.

4.2.1 Variance Model

In the case of heteroscedasticity it is profitable to plot the experimental variances s^2 or the standard deviations s versus the concentration c and to calculate by regression the relative model. The availability of the relationship s^2/c will be very useful in the weighted regression analysis, in the calculation of detection and quantification limits and in general in the operation of inverse regression (see Section 4.3.1.2). Different types of s^2 function of concentration can be evaluated. Common models are: s^2 depending on c via a linear or a quadratic model; s^2 changing with c in an exponential way.

A recent paper[7] claims, on the basis of numerous experiments in atomic absorption spectroscopy, GC/MS and inductively coupled plasma mass spectrometry, that the measurement error of an analytical method is of two types: constant at very low concentration; proportional to the concentration at higher concentration. Consequently the variance of the response is represented as $s^2 = a_0 + a_1c^2$.

4.3 CALIBRATION MODELS

The more popular model is always the straight line independent of scedasticity. Sometimes nonlinear calibration relationships have to be used, in particular when a wide dynamic range is explored. In this case, simple quadratic functions are completely satisfactory. Depending on the test of scedasticity the unweighted or the weighted regression procedure is followed.

4.3.1 Unweighted Regression

In this section we describe a least squares univariate regression analysis of response y on the analyte concentration x using a linear, first-order model (straight line):

$$y = \beta_0 + \beta_1 x + \varepsilon \tag{4.1}$$

and a linear, second-order model (quadratic model):

$$y = \beta_0 + \beta_1 x + \beta_2 x^2 + \varepsilon \tag{4.2}$$

The independent variable x is assumed unaffected by error, β_0, β_1, and β_2 are the parameters of the model and ε represents a normally distributed random error, with mean zero and constant variance $\sigma^2, \varepsilon \sim N(0, \sigma^2)$. The signal, therefore, is thought to be composed of a deterministic component predictable by the model and a random component ε. Figure 4.1 illustrates the assumptions concerning ε in the case of a straight line. The β parameters are unknown and the least squares regression furnishes their estimates by using a set of experimental data points (x_i, y_i). Thus we write:

$$\hat{y} = b_0 + b_1 x \tag{4.1'}$$

or

$$\hat{y} = b_0 + b_1 x + b_2 x^2 \tag{4.2'}$$

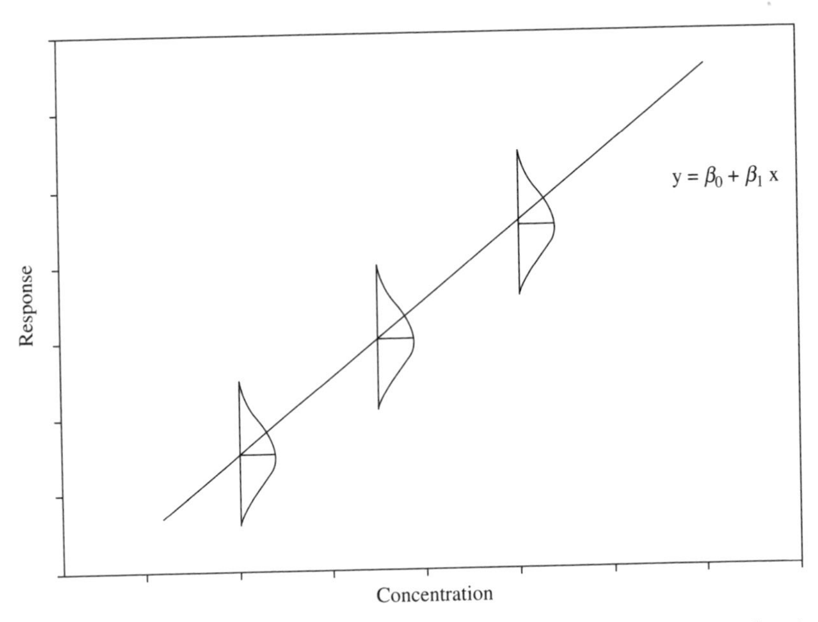

Figure 4.1 Model for linear regression in the homoscedastic case, showing distributions of measurements around the true calibration line

where $\hat{y}$ represents the predicted response of y for a given x.

It is common practice to use software packages to calculate estimates of the parameters and any other statistics of interest but we here report also the explicit formulas for any cases treated.

4.3.1.1 *Straight Line Calibration Curve*

The least squares estimates of parameters and of their variances in Equation (4.1′) are given by:

$$b_0 = \bar{y} - b_1\bar{x} \tag{4.3}$$

$$b_1 = \frac{\sum_1^n {}_i(x_i - \bar{x})y_i}{\sum_1^n {}_i(x_i - \bar{x})^2} \tag{4.4}$$

$$s_{b_0}^2 = s_{y/x}^2 \left[\frac{1}{n} + \frac{\bar{x}^2}{\sum_1^n {}_i(x_i - \bar{x})^2} \right] \tag{4.5}$$

$$s_{b_1}^2 = \frac{s_{y/x}^2}{\sum_1^n {}_i(x_i - \bar{x})^2} \tag{4.6}$$

$$s_{y/x}^2 = \frac{\sum_1^n {}_i(y_i - \hat{y}_i)^2}{n-2} \tag{4.7}$$

where n is the overall data point number:

$$n = \sum_1^k {}_j m_j$$

k is the number of the concentration levels and m_j is the number of replicates at the level j

$$\bar{x} = \frac{\sum_1^n {}_i x_i}{n}$$

$$\bar{y} = \frac{\sum_1^n {}_i y_i}{n}$$

and $s^2_{y/x}$ is the residual variance of the regression. It represents an estimate of the error variance σ^2 if the model is correct.

The adequacy of the model can be tested in several ways:[16]

(i) by the evaluation of the correlation coefficient

$$r = \frac{n\Sigma xy - \Sigma x \Sigma y}{[n\Sigma x^2 - (\Sigma x)^2]^{\frac{1}{2}}[n\Sigma y^2 - (\Sigma y)^2]^{\frac{1}{2}}};$$

(ii) by the use of the *lack of fit* test;

(iii) by inspection of the behaviour of the residuals versus concentration.

The first procedure is to be discouraged as it can lead to misinterpretation, the second one requires a good number of data points; the third is graphical in nature, easy to do and very revealing as to whether the assumptions on the errors ε and the model are correct.

4.3.1.2 *Discrimination*

The analytical application of the calibration curve is always the discrimination, i.e. obtaining x from an analytical response y, $x = (y - b_0)/b_1$. The degree of uncertainty on x depends on two factors: the uncertainty associated with the estimates b_0 and b_1, which implies the nonuniqueness of the regression line; and the uncertainty associated with the experimental response reading. A common way to take into account these two sources of error for x is the use of the $(1 - \alpha)100\%$ two-sided prediction intervals whose limits are:

$$y^{\pm} = b_0 + b_1 x \pm t_{(1-\alpha/2,\, n-2)} s_{y/x} \left[1 + \frac{1}{n} + \frac{(x - \bar{x})^2}{\sum_1^n {}_i(x_i - \bar{x})^2} \right]^{1/2} \qquad (4.8)$$

where $t_{(1-\alpha/2,n-2)}$ is $(1 - \alpha/2)100$ percentage point of Student's t-distribution on $n - 2$ degrees of freedom.[17] Equation (4.8) refers to a single y reading but, if the average $\bar{y}_m$ of m replicates $y_1, y_2, \ldots, y_m$ is considered, it changes into:

$$\bar{y}^{\pm} = b_0 + b_1 x \pm t_{(1-\alpha/2,\, n-2)} s_{y/x} \left[\frac{1}{m} + \frac{1}{n} + \frac{(x - \bar{x})^2}{\sum_1^n {}_i(x_i - \bar{x})^2} \right]^{1/2} \qquad (4.9)$$

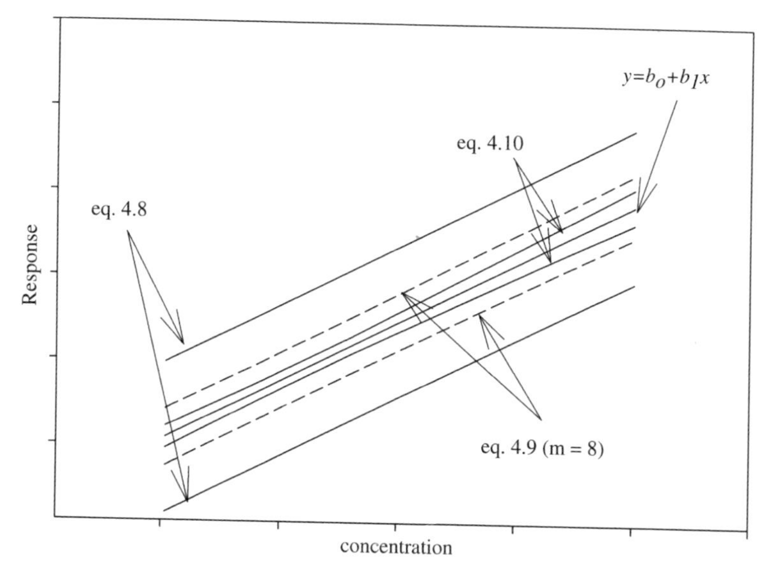

Figure 4.2 Ordinary least squares regression analysis: the middle line is the calibration line, the bounding lines are the regression bands (plots of Equation 4.10) and prediction intervals (plots of Equation 4.8 and Equation 4.9 with $m = 8$)

When m is very large, $m \to \infty$, Equation (4. 9) collapses to:

$$y^{\pm} = b_0 + b_1 x \pm t_{(1-\alpha/2,n-2)} s_{y/x} \left[\frac{1}{n} + \frac{(x-\bar{x})^2}{\sum_1^n i(x_i - \bar{x})^2} \right] \qquad (4.10)$$

which is usually called the $(1-\alpha)100\%$ *regression band*.

The real significance of Equations (4.8)–(4.10) is as follows: at a fixed x, the intervals whose limits are given by these equations contain the response y, $\bar{y}_m$, $\bar{y}_{m\to\infty}$, respectively, with probability $1-\alpha$. Figure 4.2 shows the prediction and the regression bands relative to the three cases illustrated. At this point, given an y_0 value, we are able to calculate the corresponding x_0 and its limits of confidence x_0^-, x_0^+. In Figure 4.3 two graphical procedures used to calculate x_0, x_0^- and x_0^+ are shown. In the first case, the discrimination interval (x_0^-, x_0^+) is obtained by intersecting the two-sided prediction interval with a straight line $y = y_0$ (Figure 4.3a). This definition of the interval relies on the fact that we accept as limits for x_0, values of x whose relevant predicted values exhibit

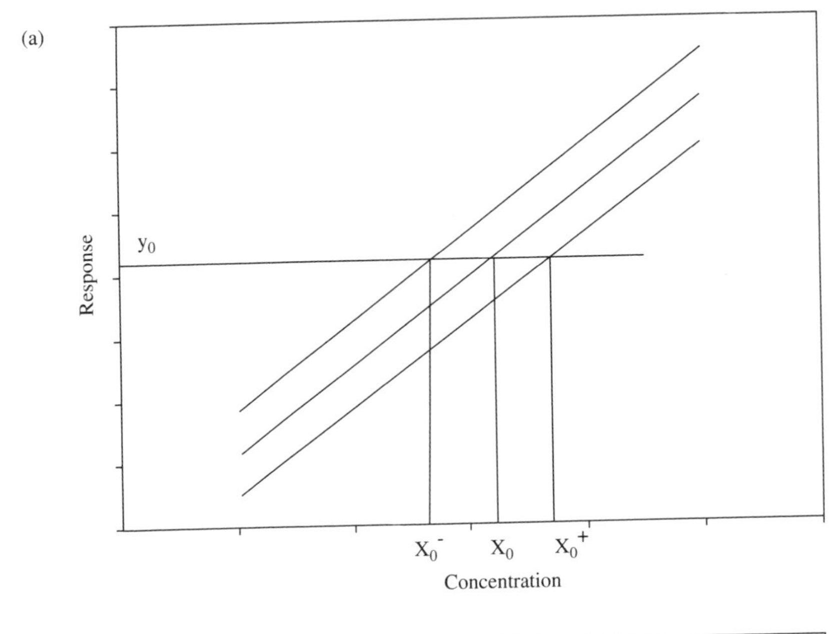

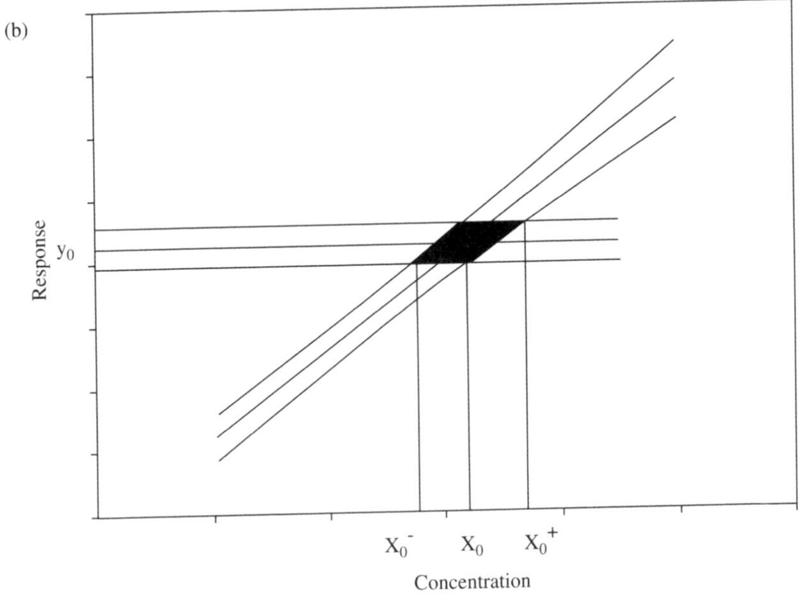

Figure 4.3 (a) The $y = y_0$ line comprises the graphical solution for confidence limits of the discriminated value x_0 by intersecting the prediction intervals. (b) A confidence interval on the response axis is converted into a discrimination interval on the x axis by observing where it intersects the confidence band on the regression line

$(1-\alpha)100\%$ prediction intervals with the experimental response y_0 as the upper and the lower limits. Alternatively in Figure 4.3(b) the $(1-\alpha)100\%$ confidence interval for x_0 is obtained by intersecting the $(1-\alpha/2)100\%$ confidence interval of y_0 with the two-sided $(1-\alpha/2)100\%$ regression band (Equation 4.10) on $y = b_0 + b_1 x$ and projecting the intersections onto the x axis.[18] The basic idea of this second approach is that all the points of the shaded area have coordinates belonging jointly to the confidence intervals of y_0 and of the regression line. In both cases the graphical procedures can be replaced by the analytical approach.

Referring to Figure 4.3(a), the limits x_0^-, x_0^+ are obtained as solution of the equations:

$$y_0 = b_0 + b_1 x_0^- + t_{(1-\alpha/2,\, n-2)} s_{y/x} \left[\frac{1}{m} + \frac{1}{n} + \frac{(x_0^- - \bar{x})^2}{\sum_1^n {}_i (x_i - \bar{x})^2} \right]^{1/2} \tag{4.11a}$$

$$y_0 = b_0 + b_1 x_0^+ - t_{(1-\alpha/2,\, n-2)} s_{y/x} \left[\frac{1}{m} + \frac{1}{n} + \frac{(x_0^+ - \bar{x})^2}{\sum_1^n {}_i (x_i - \bar{x})^2} \right]^{1/2} \tag{4.11b}$$

Referring to Figure 4.3(b), the limits x_0^-, x_0^+ are calculated using the equations:

$$y_0 + t_{(1-\alpha/4,\, m-1)} s_{y_0} = b_0 + b_1 x_0^+ - t_{(1-\alpha/4,\, n-2)} s_{y/x} \left[\frac{1}{n} + \frac{(x_0^+ - \bar{x})^2}{\sum_1^n {}_i (x_i - \bar{x})^2} \right]^{1/2} \tag{4.12a}$$

$$y_0 - t_{(1-\alpha/4,\, m-1)} s_{y_0} = b_0 + b_1 x_0^- + t_{(1-\alpha/4,\, n-2)} s_{y/x} \left[\frac{1}{n} + \frac{(x_0^- - \bar{x})^2}{\sum_1^n {}_i (x_i - \bar{x})^2} \right]^{1/2} \tag{4.12b}$$

where s_{y_0} is the experimental standard deviation of the mean y_0 of m responses. Obviously the value of s_{y_0} must be close to $s_{y/x}/m$.

4.3.1.3 *Detection and Quantification Limits*

The availability of the calibration line and of related prediction functions coming from a well suited calibration design permits the calculation of the critical, detection and quantification limits according to a well established procedure.[8,19] Following the current literature[1,3,8] the critical level L_C is the assay signal above which a response is reliably attributed to the presence of analyte at concentration x_C; the detection limit L_D is the signal corresponding to an analyte concentration level x_D, also called the minimum detectable concentration, which may be a priori expected to be recognized; finally, the quantification limit L_Q is, in response units, a signal whose precision satisfies an expected value. The availability of the calibration function permits the immediate transfer from the signal to the concentration domain for the above described limits (Figure 4.4). On the basis of statistical considerations the critical level is defined by the rejection of the null hypothesis H_0: concentration = 0 at the critical level α (for example, $\alpha = 0.05$) (type I error rate). Referring to Figure 4.4 we mean that the intercept, i.e. the signal at concentration equal to zero, has only 5% probability to overcome the critical level L_C, so that any response greater than L_C is reliably attributed to the presence of analyte. For L_D, a type II error or

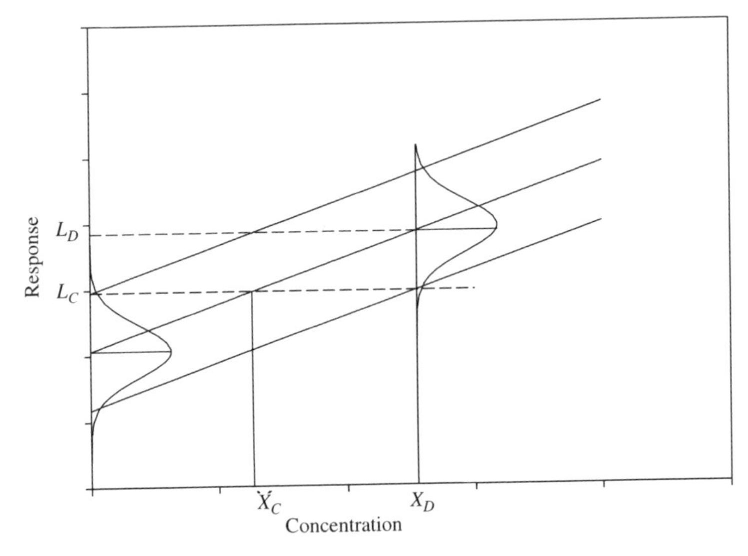

Figure 4.4 The critical level (x_C, L_C) and the detection level (x_D, L_D). The figure includes the calibration curve with confidence intervals for a predicted observation

false negative error, must be invoked. Namely, a concentration x_C gives a mean response equal to L_C but a single response is found over L_C only with a probability of 50%; therefore the false negative error rate β is equal to 50%. Following these considerations the detection limit L_D must be defined controlling the β value which usually is set equal to α. The x_D value is the abscissa of the intersection of a parallel line to the x axis passing through L_C with the lower one-sided $(1-\beta)100\%$ prediction function.

The quantification limit L_Q can be defined in different ways:

(i) $$\frac{L_Q - b_0}{s_{L_C}} = 10$$

a net response equal to ten times the standard deviation at the lowest detectable signal L_C;[8]

(ii) $$\frac{L_Q}{s_{y/x}\left[1+\frac{1}{n}+\frac{(x_Q-\bar{x})^2}{\Sigma(x_i-\bar{x})^2}\right]^{1/2}} = 10$$

a signal ten times the standard deviation of the prediction value at the concentration x_Q;[3]

(iii) $$\frac{L_Q - b_0}{s_{b_0}} = 10$$

a net signal equal to ten times the standard deviation of the intercept of the calibration curve.[6]

4.3.1.4 *Quadratic Calibration Curve*

It is common practice that techniques used over a wide concentration range, such as GC/MS, can exhibit curvilinear calibration plots fitted by the equation:

$$\hat{y} = b_0 + b_1 x + b_2 x^2$$

By simple arrangement the model changes into:

$$\hat{y} = \bar{y} + b_1(x-\bar{x}) + b_2(x^2 - \overline{x^2})$$

where

$$\bar{y} = \frac{\sum_1^n {}_i y_i}{n} \qquad \bar{x} = \frac{\sum_1^n {}_i x_i}{n} \qquad \overline{x^2} = \frac{\sum_1^n {}_i x^2}{n}$$

and n is the overall number of calibration points.
The parameters b_1 and b_2, estimated by least-squares procedure, are:

$$b_1 = \frac{s_{ff}s_{xy} - s_{fx}s_{fy}}{\Delta}$$
$$b_2 = \frac{s_{xx}s_{fy} - s_{fx}s_{xy}}{\Delta}$$

where

$$\Delta = s_{xx}s_{ff} - s_{fx}^2$$
$$s_{xx} = \sum_1^n {}_i x_i^2 - n\bar{x}^2$$
$$s_{fx} = \sum_1^n {}_i x_i^3 - n\bar{x}\overline{x^2}$$
$$s_{ff} = \sum_1^n {}_i x_i^4 - n(\overline{x^2})^2$$
$$s_{xy} = \sum_1^n {}_i x_i y_i - n\bar{x}\bar{y}$$
$$s_{fy} = \sum_1^n {}_i x_i^2 y_i - n\overline{x^2}\bar{y}.$$

Regression analysis also gives formulas for the coefficient variances and covariances:

$$s_{b_1}^2 = \frac{s_{ff}}{\Delta} s_{y/x}^2$$
$$s_{b_2}^2 = \frac{s_{xx}}{\Delta} s_{y/x}^2$$
$$s_{b_1,b_2}^2 = -\frac{s_{fx}}{\Delta} s_{y/x}^2$$
$$s_{b0}^2 = \left(\frac{1}{n} + \frac{\bar{x}^2 s_{ff}}{\Delta} + \frac{\overline{x^2}^2 s_{xx}}{\Delta} - \frac{2\bar{x}\overline{x^2} s_{fx}}{\Delta}\right) s_{y/x}^2$$

where

$$s_{y/x}^2 = \frac{\sum_1^n {}_i(y_i - \hat{y}_i)^2}{n-3}$$

estimates the measurement population variance σ^2.

Following the same arguments reported in Section 4.3.1.2, Equations (4.8)–(4.10) for the prediction and regression bands now become:

$$y^{\pm} = \bar{y} + b_1(x-\bar{x}) + b_2(x^2 - \overline{x^2}) \pm t_{(1-\alpha/2,n-3)} s_{y/x}[1+U(x)]^{1/2} \quad (4.13)$$

$$\bar{y}_m^{\pm} = \bar{y} + b_1(x-\bar{x}) + b_2(x^2 - \overline{x^2}) \pm t_{(1-\alpha/2,n-3)} s_{y/x}\left[\frac{1}{m} + U(x)\right]^{1/2} \quad (4.14)$$

$$\bar{y}_{m\to\infty}^{\pm} = \bar{y} + b_1(x-\bar{x}) + b_2(x^2 - \overline{x^2}) \pm t_{(1-\alpha/2,n-3)} s_{y/x}[U(x)]^{1/2} \quad (4.15)$$

where

$$U(x) = \frac{1}{n} + (x-\bar{x})^2 \frac{s_{ff}}{\Delta} + (x^2 - \overline{x^2})^2 \frac{s_{xx}}{\Delta} - 2(x-\bar{x})(x^2 - \overline{x^2}) \frac{s_{fx}}{\Delta}.$$

Figure 4.5 shows a three-parameter parabolic calibration function together with the two-sided 95% regression band and the associated two-sided prediction functions.

To discriminate an unknown x_0 value with its confidence limits from an experimental y_0 response and to calculate x_C, x_D, and x_Q the same guidelines used for the straight line model are to be followed. The more cumbersome mathematics involved in the analytical calculation can be easily overcome resorting to graphical solution.

4.3.2 Weighted Regression

When repeated measurements for each concentration level are available the scedasticity of the data can be evaluated. It is known that disregarding the heteroscedasticity slightly affects the estimates of the parameters but heavily influences confidence and detection limits. The successful approach in treating this case is weighted regression analysis. The data are weighted so that the homoscedasticity is obtained in the transformed data. Each measurement y_i at concentration level $x_i (i = 1, 2, \ldots, n)$ is multiplied with the weighting factor $w_i^{1/2} = \sigma/\sigma_i$, where σ_i^2 is the variance of the replicate responses obtained at the concentration level

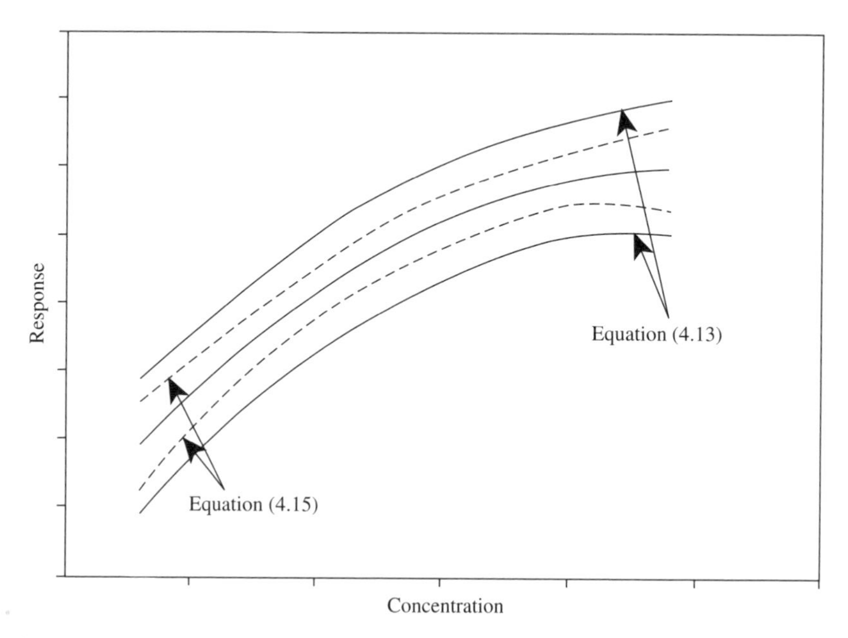

Figure 4.5 OLS regression analysis: a three-parameter parabolic calibrating function (middle line) with its regression bands (dashed curve) and prediction functions

x_i and σ^2 is the unknown common variance of all the weighted measurements.

4.3.2.1 Weighted Straight Line Calibration Curve

The WLS estimates b_{0w} and b_{1w} of the parameters β_0 and β_1 in Equation (4.1) and their variances are given by the following formulas:

$$b_{0w} = \bar{y}_w - b_{1w}\bar{x}_w \tag{4.16}$$

$$b_{1w} = \frac{\sum_1^n {}_i w_i(x_i - \bar{x}_w)y_i}{\sum_1^n {}_i w_i(x_i - \bar{x}_w)^2} \tag{4.17}$$

$$s_{b_{0w}}^2 = \left(\frac{1}{\sum w_i} + \frac{\bar{x}_w^2}{\sum w_i(x_i - \bar{x}_w)^2}\right)(s_{y/x})_w^2 \tag{4.18}$$

$$s_{b_{1w}}^2 = \frac{1}{\sum w_i(x_i - \bar{x}_w)^2}(s_{y/x})_w^2 \tag{4.19}$$

$$(s_{y/x})_w^2 = \frac{\sum w_i(y_i - \hat{y}_{iw})^2}{n-2} \tag{4.20}$$

where $\hat{y}_{iw} = b_{0w} + b_{1w}x_i$ is the predicted response at level x_i

$$\bar{x}_w = \frac{\sum w_i x_i}{\sum w_i}$$

and

$$\bar{y}_w = \frac{\sum w_i y_i}{\sum w_i}.$$

A residual standard deviation near unity, $(s_{y/x})_w \approx 1^5$ is obtained using the inverse variance $1/s_i^2$ as w_i at each concentration level x_i.

The values of the variances s_i^2, i.e. of the weights, can be estimated by modelling the variance or the standard deviation as a function of concentration (see Section 4.2.1).

A two-sided $(1 - \alpha)100\%$ weighted prediction interval around a predicted response $\hat{y}_{iw}$ at concentration x_i is calculated as:

$$y_{iw}^{\pm} = \hat{y}_{iw} \pm t_{(1-\alpha/2,n-2)}(s_{y/x})_w \left[\frac{1}{w_i} + U_w(x_i)\right]^{1/2} \tag{4.21}$$

where

$$U_w(x) = \frac{1}{\sum w_i} + \frac{(x - \bar{x}_w)^2}{\sum w_i (x_i - \bar{x}_w)^2}.$$

We remark again that the first term inside the parentheses is the variance of a single response at x_i.

Therefore this term changes into $1/mw_i$ when we predict a mean of m responses. For $m \to \infty$ this term vanishes and Equation (4.21) then describes the regression band of a WLS straight line. Figure 4.6 shows a quadratic calibration curve together with the regression band and the two-sided prediction curves.

4.3.2.2 *Discrimination and Detection Limits*

The same arguments developed in Sections 4.3.1.2 and 4.3.1.3 hold. To discriminate an unknown x_0 value and its relative confidence limits x_0^- and x_0^+ from a known response y_0, the two procedures above described

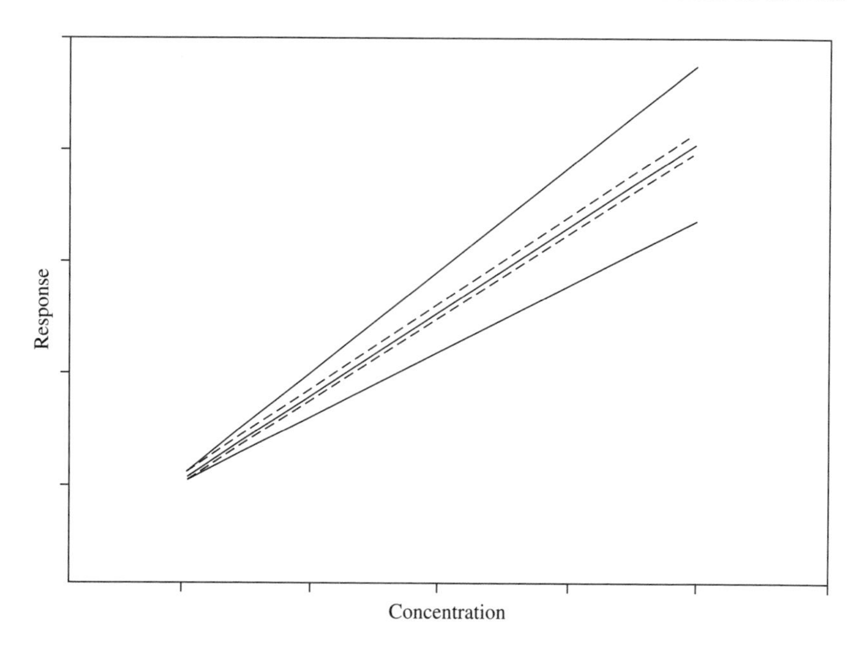

Figure 4.6 WLS regression analysis: straight line calibrating function and its regressions bands (dashed line) and prediction functions

can be followed. The first approach gives:

$$y_0 = b_{0w} + b_{1w}x_0^- + t_{(1-\alpha/2,n-2)}(s_{y/x})_w \left[\frac{1}{mw_{x_0^-}} + U_w(x_0^-)\right]^{1/2} \quad (4.22a)$$

and

$$y_0 = b_{0w} + b_{1w}x_0^+ - t_{(1-\alpha/2,n-2)}(s_{y/x})_w \left[\frac{1}{mw_{x_0^+}} + U_w(x_0^+)\right]^{1/2} \quad (4.22b)$$

where the weights $w_{x_0^-}$ and $w_{x_0^+}$ are obtained from the model of the standard deviation or of the variance. The limits x_0^- and x_0^+ are obtained by solving a quadratic equation or by an easy graphical procedure. The second procedure gives:

$$y_0 + t_{(1-\alpha/4,m-1)}s_{y0} = b_{0w} + b_{1w}x_0^+ - t_{(1-\alpha/4,n-2)}(s_{y/x})_w U_w(x_0^+)^{1/2} \quad (4.23a)$$

and

$$y_0 - t_{(1-\alpha/4,m-1)} s_{y0} = b_{0w} + b_{1w} x_0^- + t_{(1-\alpha/4,n-2)} (s_{y/x})_w U_w(x_0^-)^{1/2} \quad (4.23b)$$

where the term s_{y_0} is the experimental standard deviation of the mean y_0 of m measurements. Again the analytical or the graphical procedure furnishes the x_0^- and x_0^+ limits.

Equation (4.21), calculated at $x_i = 0$ using the $(1-\alpha)100$ percentage point of Student's t-distribution (one-sided interval), determines y_w^+ which is assumed as the critical signal L_{Cw} and then:

$$x_{Cw} = \frac{L_{Cw} - b_{0w}}{b_{1w}} \quad (4.24)$$

Obtaining of L_{Cw} requires the insertion of an estimate of the weight w_0 at zero concentration. Two approaches can be followed: the use of the weight at the lowest spiking concentration[3] or the extrapolated value obtained with the model of the variance.[8] Intersection of the parallel to the abscissa at the level L_{Cw} with the lower $(1-\beta)100\%$ one-sided prediction function gives the detection limit x_{Dw}. The graphical solution is immediate; otherwise a more cumbersome iterative approach requiring the correct value for w_{x_D} can be used.

For the calculation of x_{Qw} the same definitions reported in Section 4.3.1.3 hold. In the WLS approach the starting equations are:

$$\frac{L_{Qw} - b_{0w}}{s_{L_{cw}}} = 10 \quad (4.25a)$$

$$\frac{L_{Qw}}{(s_{y/x})_w \left[\frac{1}{w_{x_{Qw}}} + \frac{1}{\Sigma w_i} + \frac{(x_{Qw} - \bar{x}_w)^2}{\Sigma w_i (x_i - \bar{x}_w)^2}\right]^{1/2}} = 10 \quad (4.25b)$$

$$\frac{L_{Qw} - b_{0w}}{s_{b_{0w}}} = 10 \quad (4.25c)$$

4.3.2.3 *Weighted Quadratic Calibration Curve*

In the case of a quadratic calibration curve with nonuniform variance the WLS estimates of the parameters b_{1w} and b_{2w} of the model

$$\hat{y}_w = \bar{y}_w + b_{1w}(x - \bar{x}_w) + b_{2w}(x^2 - \overline{x^2}_w) \quad (4.26)$$

are

$$b_{1w} = \frac{s_{ff}s_{xy} - s_{fx}s_{fy}}{\Delta}$$
$$b_{2w} = \frac{s_{xx}s_{fy} - s_{fx}s_{xy}}{\Delta}$$

where

$$\Delta = s_{xx}s_{ff} - s_{fx}^2$$
$$s_{xx} = \sum_1^n {}_iw_i x_i^2 - \left(\sum_1^n {}_iw_i\right)\bar{x}_w^2$$
$$s_{fx} = \sum_1^n {}_iw_i x_i^3 - \left(\sum_1^n {}_iw_i\right)\bar{x}_w\overline{x^2}_w$$
$$s_{ff} = \sum_1^n {}_iw_i x_i^4 - \left(\sum_1^n {}_iw_i\right)(\overline{x^2}_w)^2$$
$$s_{xy} = \sum_1^n {}_iw_i x_i y_i - \left(\sum_1^n {}_iw_i\right)\bar{x}_w\bar{y}_w$$
$$s_{fy} = \sum_1^n {}_iw_i x_i^2 y_i - \left(\sum_1^n {}_iw_i\right)\overline{x^2}_w\bar{y}_w$$

n is the overall number of calibration points

$$\bar{x}_w = \frac{\sum_1^n {}_iw_i x_i}{\sum_1^n {}_iw_i}$$
$$\overline{x^2}_w = \frac{\sum_1^n {}_iw_i x^2}{\sum_1^n {}_iw_i}$$

and

$$\bar{y}_w = \frac{\sum_1^n {}_iw_i y_i}{\sum_1^n {}_iw_i}.$$

The coefficient variances and covariances are:

$$s_{b_{1w}}^2 = \frac{s_{ff}}{\Delta}(s_{y/x}^2)_w$$
$$s_{b2w}^2 = \frac{s_{xx}}{\Delta}(s_{y/x}^2)_w$$
$$s_{b_1,b_2w}^2 = -\frac{s_{fx}}{\Delta}(s_{y/x}^2)_w$$

where

$$(s^2_{y/x})_w = \frac{\sum_1^n {}_i w_i (y_i - \hat{y}_{iw})^2}{n-3}. \tag{4.27}$$

Equations (4.13)–(4.15) now become:

$$y^{\pm}_{iw} = \hat{y}_{iw} \pm t_{(1-\alpha/2, n-3)} (s_{y/x})_w \left[\frac{1}{w_i} + U_w(x_i)\right]^{1/2} \tag{4.28}$$

$$(\bar{y}_m)^{\pm}_{iw} = \hat{y}_{iw} \pm t_{(1-\alpha/2, n-3)} (s_{y/x})_w \left[\frac{1}{m w_i} + U_w(x_i)\right]^{1/2} \tag{4.29}$$

$$(\bar{y}_{m\to\infty})^{\pm}_{iw} = \hat{y}_{iw} \pm t_{(1-\alpha/2, n-3)} (s_{y/x})_w [U_w(x_i)]^{1/2} \tag{4.30}$$

where $\bar{y}_m$ is the average of m replicated responses at x_i, and

$$U_w(x) = \frac{1}{\sum_1^n {}_i w_i} + (x_i - \bar{x}_w)^2 \frac{s_{ff}}{\Delta} + (x_i^2 - \overline{x^2}_w)^2 \frac{s_{xx}}{\Delta} - 2(x_i - \bar{x}_w)(x_i^2 - \overline{x^2}_w) \frac{s_{fx}}{\Delta}.$$

Figure 4.7 shows a quadratic calibration curve together with the regression band and the two-sided prediction curves.

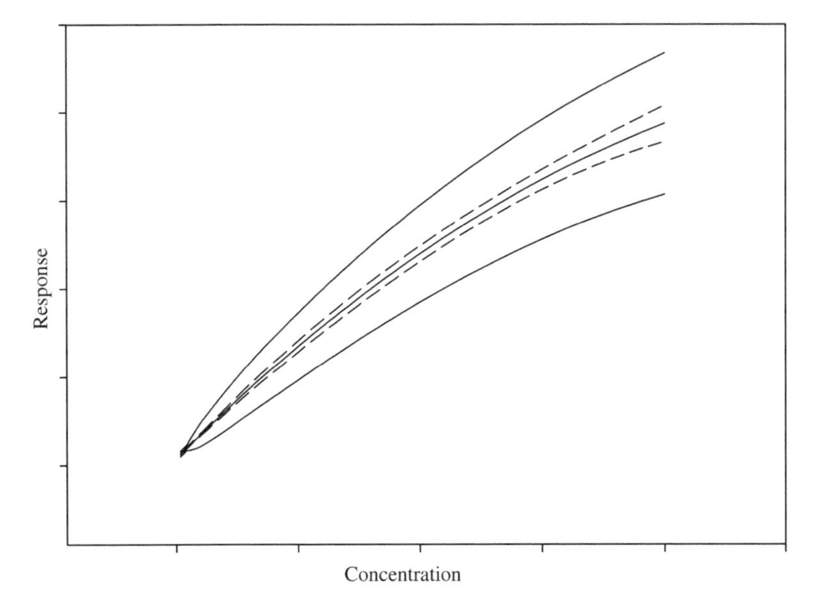

Figure 4.7 WLS regression analysis: a three-parameter parabolic calibrating function (middle line) with its regression bands (dashed curve) and prediction functions

To find a discriminated x_0 value together with its confidence interval and to calculate the critical value L_{Cw}, the detection L_{Dw} and the quantification L_{Qw} limits the arguments and the approaches described in Sections 4.3.1.2 and 4.3.1.3 hold.

4.3.3 A Practical Example

The construction of a calibration curve for the GC/MS measurement of chloroform in groundwater is illustrated using the approach outlined above.

The measurement data were obtained by an analytical procedure which mostly followed the recommendations of the Environmental Protection Agency (EPA) method.[20] The solution analysis was performed using a quadrupole GC/MS instrument equipped with purge-and-trap cryomodule.

The concentration levels ranged from zero to 4 μg/L, defining a concentration interval which is used both for establishing the detection limit and for discriminating an unknown sample.

For each standard concentration ten solutions were made up and analysed with the results shown in Table 4.1 and in Figure 4.8. It is evident that larger absolute errors were found at the upper extreme of the calibration region. The analysis of the plot of the residuals obtained from a preliminary regression with an unweighted straight line model (Figure 4.9) points out that the standard deviations for each set of y values are unequal and indicates that there is a light trend in the

Table 4.1 Calibration data in terms of ratios of peak area of chloroform and of IS (fluorobenzene) as a function of chloroform concentration

No. of replicates	Concentration level (μg/L)								
	0	0.03	0.1	0.2	0.4	0.8	1.6	3.2	4
1	0.0219	0.0487	0.1154	0.2290	0.4129	0.7597	1.4237	2.6167	3.3549
2	0.0108	0.0355	0.0922	0.1687	0.3397	0.6061	1.1379	2.1139	2.6265
3	0.0282	0.0443	0.1154	0.1861	0.3604	0.6668	1.2830	2.3026	2.8001
4	0.0261	0.0519	0.0985	0.1895	0.3541	0.6317	1.2226	2.1846	2.6932
5	0.0369	0.0394	0.0807	0.1694	0.3596	0.6315	1.1581	2.1704	2.7554
6	0.0148	0.0554	0.1193	0.2020	0.3960	0.7103	1.2745	2.3174	2.8915
7	0.0122	0.0545	0.1371	0.2126	0.3838	0.7169	1.3113	2.5015	3.0139
8	0.0308	0.0526	0.1098	0.2543	0.3859	0.6781	1.2110	2.1774	2.7617
9	0.0261	0.0510	0.1224	0.2441	0.3989	0.6909	1.4272	2.5282	3.0527
10	0.0346	0.0278	0.0780	0.1475	0.2816	0.5494	1.0956	1.9437	2.4083

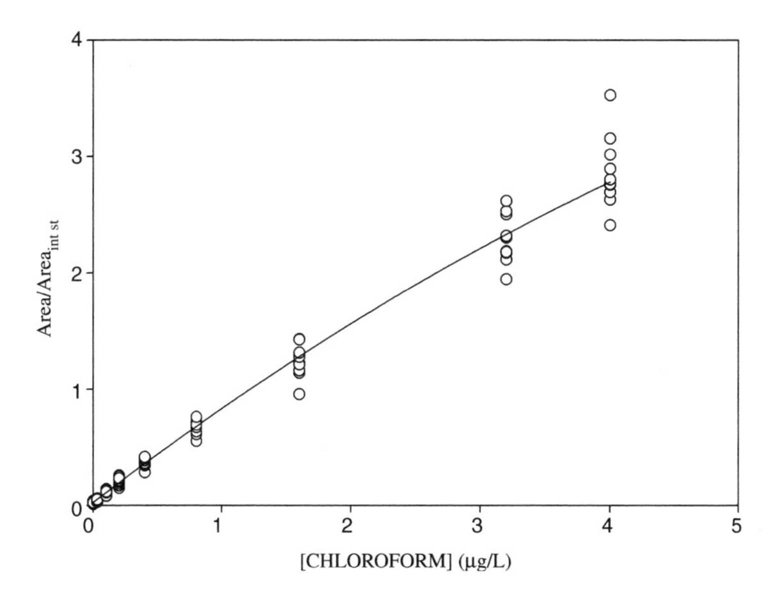

Figure 4.8 (○) Experimental ratios of the peak area of chloroform and of IS (fluorobenzene) as a function of chloroform concentration. (—) Calibration function estimated by weighted regression: $y = 0.0260 + 0.8445x - 0.0391x^2$

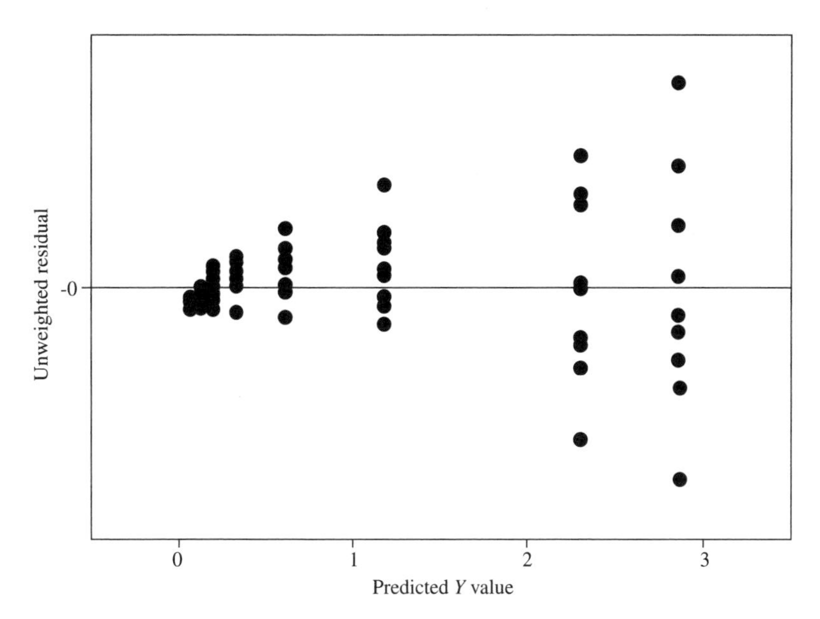

Figure 4.9 Residual plot after unweighted regression with a straight line model of the chloroform calibration data. OLS calibration curve: $y = 0.0613 + 0.702x$ ($r = 0.9924$)

residuals. The former observation suggests the use of weighting in order to take greater account of the more precisely defined data; the latter indicates that a curved calibration line is preferable to a straight line.

The use of the inverse of the variance of the ratios of the MS responses at each concentration level as the weighting factor appears suitable for obtaining a reliable calibration and establishing a reliable detection limit. Actually, in spite of the wide concentration range used, the proposed choice of the weighting factors forces the regression curve to pass more closely to the points near the origin. In this way, the experimental design chosen agrees with the requirements for the achievement of a reliable detection limit.

Table 4.2 reports the results obtained using a straight line and a quadratic model in a weighted regression approach. The weighting factors were calculated from the variance model $V = V_0 + V_1C^2$, where C is the concentration level, whose parameters $V_0 = 3.060 \times 10^{-4}$ and $V_1 = 5.536 \times 10^{-3}$ were determined by the fit of the experimental variances (Figure 4.10).

The adequacy of the quadratic model for the calibration curve can be deduced from the plot of the weighted residuals versus the predicted y values, as shown in Figure 4.11. The residuals appear randomly

Table 4.2 Practical example: regression parameters for the weighted straight line $y = \beta_{0w} + \beta_{1w}x$ and the quadratic model $y = \beta_{0w} + \beta_{1w}x + \beta_{2w}x^2$, as calibration curves for the data reported in Table 4.1, and critical limit x_{Cw}, detection limit x_{Dw} and discriminated x_0 value, together with its 95% confidence limits x_0^- and x_0^+

Model	Parameter	Notation	Estimate
Straight line	Slope	b_{1w}	0.0357
	Intercept	b_{0w}	0.7527
	Residual standard deviation	$(s_{y/x})_w$	1.080
	Critical limit ($\alpha = 0.05$)	x_{Cw}	0.05
	Detection limit ($\alpha = 0.05, \beta = 0.05$)	x_{Dw}	0.11
	Discriminated value for $y_{0m} = 0.66$[a]	x_0	0.83[b]
	95% confidence limits for x_0	(x_0^-, x_0^+)	(0.78, 0.90)[c]
Quadratic	First coefficient	b_{2w}	−0.0391
	Second coefficient	b_{1w}	0.8445
	Intercept	b_{0w}	0.0260
	Residual standard deviation	$(s_{y/x})_w$	0.939
	Critical limit ($\alpha = 0.05$)	x_{Cw}	0.04
	Detection limit ($\alpha = 0.05, \beta = 0.05$)	x_{Dw}	0.08
	Discriminated value for $y_0 = 0.66$[a]	x_0	0.78[b]
	95% confidence limits for x_0	(x_0^-, x_0^+)	(0.74, 0.83)[c]

[a] Mean of $m = 10$ response values.
[b] Nominal x value equal to 0.8.
[c] Straight line: from Equations 4.11a and 4.11b; quadratic: from Equation (4.22a) and (4.22b).

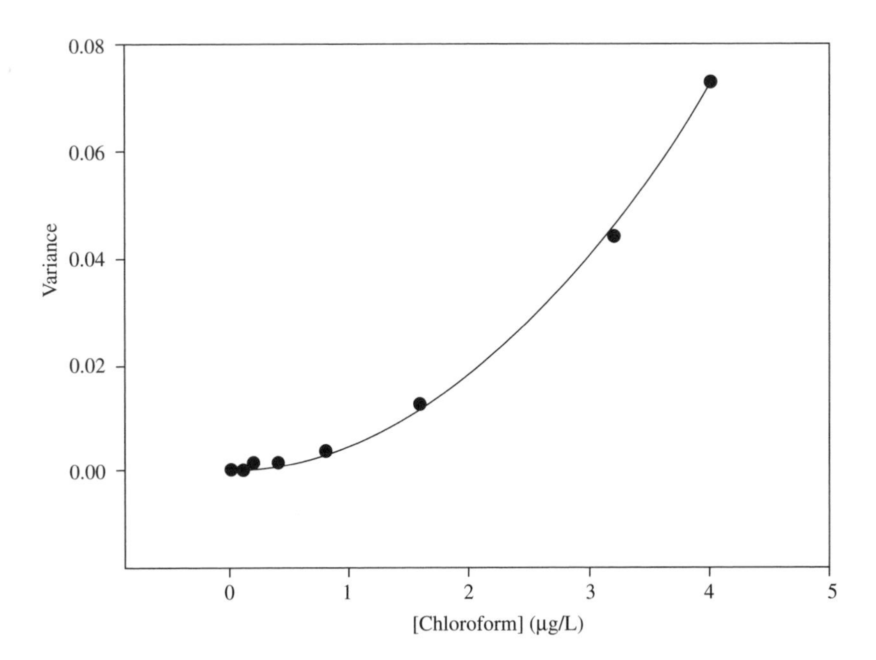

Figure 4.10 Plot of the experimental variances (•) of response against concentration for the data in Table 4.1 and corresponding fitting curve (—) obtained by the relationship $y = 0.0004725 + 0.004492x^2$

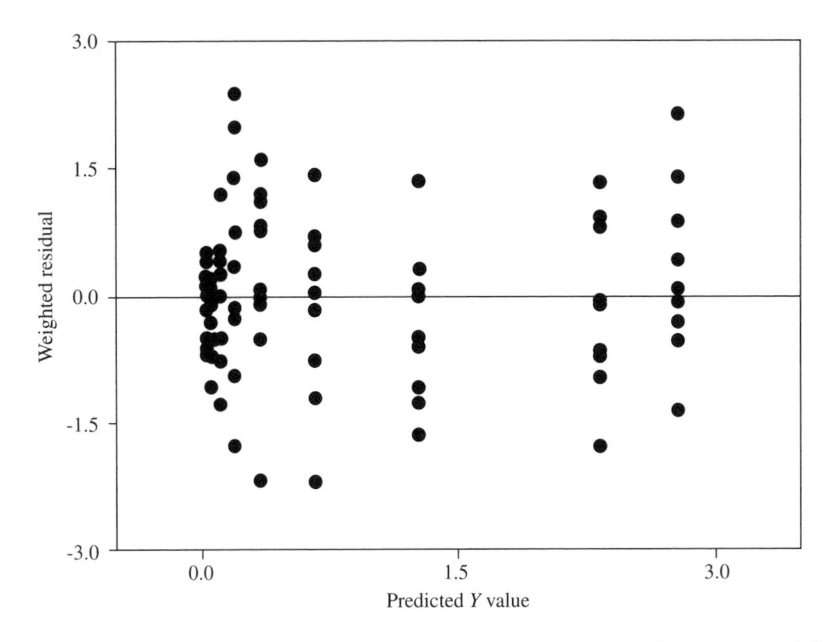

Figure 4.11 Residual plot after weighted regression with a quadratic curve of the chloroform calibration data using the inverse of the modelled variance as weighting factor. WLS calibration curve: $y = 0.0260 + 0.8445x - 0.0391x^2$ (continuous curve in Figure 4.8)

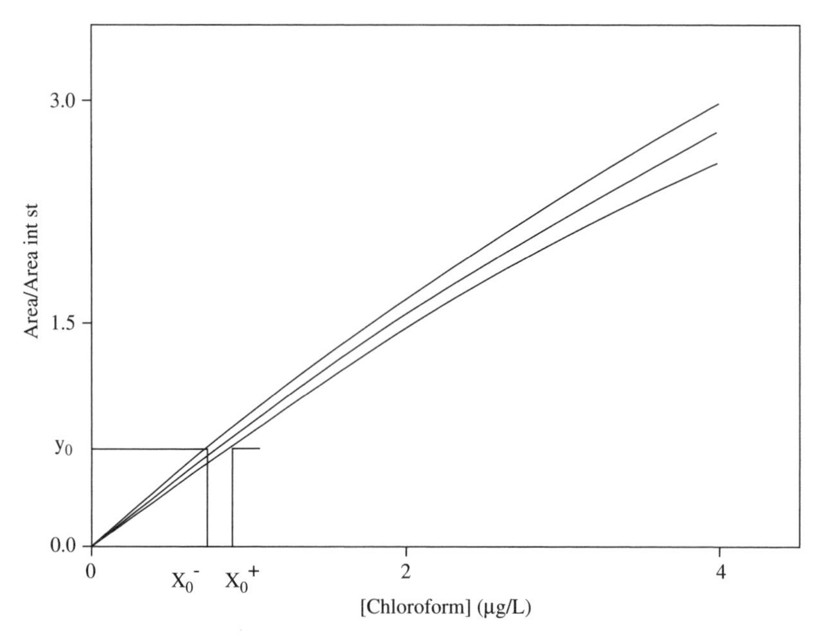

Figure 4.12 95% confidence limits for the discriminated chloroform concentration of an unknown sample whose response is the mean of ten measurements. The confidence limits are calculated as intersections of the $y_0 = 0.66$ line with the bounding lines representing two-sided 95% prediction functions ($m = 10$). The middle line is the calibration curve

distributed around zero. Further, the residual standard deviation $(s_{y/x})_w$ for the quadratic calibration curve is very close to unity, as theoretically required for a weighted good fit.

Finally, Figure 4.12 shows the quadratic calibration curve together with the 95% two-sided prediction function ($m = 10$) utilized to execute the inverse regression and to determine the confidence limits for the discriminated x value.

4.4 DIFFERENT APPROACHES TO ESTIMATE DETECTION AND QUANTIFICATION LIMITS

Other procedures reported in the literature are based: (i) again on suitable calibration designs but with considerations different from those described above; (ii) on the variability in analyte response at a single, arbitrary spiked concentration; (iii) on the signal to noise ratio where the noise magnitude is taken as an estimate of the blank standard deviation. In the first approach OLS or WLS regression furnishes, as usual, the calibration straight line with the associated regression residual standard deviation $s_{y/x}$ and the standard deviation of the intercept s_{b_0}. Then two options can be followed: the limit of detection is defined by the net response equal either to three times $s_{y/x}$ or to three times s_{b_0}. It is well known that the traditional value 3 is simply the rounding off of 3.29, i.e. two times 1.645. This choice implies α and β rates of false negative errors both equal to 5% since for a gaussian distribution the critical value of the one-tailed standardized variable z value for $\alpha = 0.05$ is indeed 1.645. It can be observed that even if not explicitly mentioned, L_C is introduced at the net signal level 1.645 times the chosen standard deviation. The limit of quantification corresponds to a net signal equal to ten times the standard deviation chosen.

In the second procedure,[21] adopted by the EPA, the limit called the *method detection limit* (*MDL*) is given by:

$$MDL = t_{(\alpha=0.01,\, n-1=6)} s = 3.14\, s$$

where s is the standard deviation of a sample of $n = 7$ replicates in which the analyte was spiked at a concentration of two to five times the suspected *MDL*, and t is the single-sided 99%, six degrees of freedom variate of Student's distribution.

Finally, in the third procedure the signal to noise ratio is considered in the signal domain. This procedure does aim not to establish whether a defined analyte concentration will be detectable but instead whether a measured signal is attributable to the presence of the analyte, i.e. to establish a critical level L_C. Two approaches are proposed on the basis of the independence (Case A) or dependence (Case B) of the measurements relevant to the blank and to the sample.

Case A. Let $\bar{x}_B$ be the mean of n_B replicated measurements of the blank, $\bar{x}_A$ the mean of n_A replicated measurements of the hypothesized analyte signal and $\overline{D} = \bar{x}_A - \bar{x}_B$. Under the assumption of $\sigma_A^2 = \sigma_B^2 = \sigma^2$ the variance of the difference is given by:

$$\sigma_{\overline{D}}^2 = \sigma^2 \left(\frac{1}{n_A} + \frac{1}{n_B} \right). \tag{4.31}$$

The insertion in Equation (4.31) of an estimate of σ^2 calculated by means of the relationship:

$$s^2 = \frac{(n_A - 1)s_A^2 + (n_B - 1)s_B^2}{n_A + n_B - 2}$$

gives

$$s_{\overline{D}}^2 = s^2\left(\frac{1}{n_A} + \frac{1}{n_B}\right).$$

The detection limit, or, better, the threshold value defined as the minimum signal to noise ratio detectable is:

$$\frac{\overline{D}}{s_{\overline{D}}} = t_{(1-\alpha,\, v)}$$

where α is the accepted rate of false positive error and $v = n_A + n_B - 2$ is the number of degree of freedom.

Three parts should be noted: a previous F-test at a chosen level is required to check the equality of s_A^2 and s_B^2; a calibration step allows the passage from the signal domain to the concentration domain; by making more measurements the critical level, so defined, can be reduced to a very small value.

Case B. In this situation it is advisable to take a measurement of the background signal $x_{B,i}$ followed by a measurement of the sample signal $x_{A,i}$. The net signal is therefore $d_i = x_{A,i} - x_{B,i}$, whose variance can be estimated as $s_d^2 = 2s_B^2$.

The mean value $\bar{d}$ of n net signals is considered significantly different from zero if:

$$\frac{\bar{d}}{\dfrac{s_d}{\sqrt{n}}} > t_{(1-\alpha,\, n-1)}$$

REFERENCES

1. L. A. Currie, *Chem. Intell. Lab. Systems*, **37**, 151–181 (1997).
2. R. D. Gibbons, *Statistical Methods For Groundwater Monitoring*, John Wiley & Sons, Inc., New York, Chapter 5 (1994).
3. L. Oppenheimer, T. P. Capizzi, R. M. Weppeiman and H. Metha, *Anal. Chem.*, **55**, 638–643 (1983).
4. C. A. Clayton, J. W. Hines and P. D. Elkin, *Anal. Chem.* **59**, 2506–2514 (1987).
5. J. Mocak, A. M. Bond, S. Mitchell and G. Scollary, *IUPAC Pure Appl. Chem.*, **69**, 298–328 (1997).
6. J. Vial and A. Jardy, *Anal. Chem.* **71**, 2672–2677 (1999).

7. M. D. Wilson, D. M. Rocke, B. Durbin and H. D. Kahn, *Anal. Chim. Acta*, **509**, 197–208 (2004).
8. M. E. Zorn, R. D. Gibbons and W. C. Sonzogni, *Anal. Chem.*, **69**, 3069–3075 (1997).
9. L. M. Schwartz, *Anal. Chem.*, **51**, 723–727 (1979).
10. L. M. Schwartz, *Anal. Chem.*, **55**, 1424–1426 (1983).
11. B. J. Millard, *Quantitative Mass Spectrometry*, Heiden, London (1978).
12. J. R. Chapman and E. Bailey, *J. Chromatogr.*, **89**, 215–224 (1974).
13. D. A. Schoeller, *Biomed. Mass Spectrom.*, **3**, 265–271 (1976).
14. J. S. Garden, D. G. Mitchell and W. N. Mills, *Anal. Chem.*, **52**, 2310–2315 (1980).
15. D. L. Massart, B. G. M. Vandeginste, S. N. Deming, Y. Michotte and L. Kaufman, *Chemometrics: a Textbook*, Elsevier, Amsterdam (1988).
16. Analytical Methods Committee, *Analyst*, **119**, 2363–2366 (1994).
17. K. A. Brownlee, *Statistical Theory and Methodology*, John Wiley & Sons, Inc., New York (1965).
18. R. G. Miller, *Simultaneous Statistical Inference*, McGraw-Hill, New York (1966).
19. A. Hubaux and G. Vos, *Anal. Chem.*, **42**, 849–855 (1970).
20. J. W. Munch (Ed.), *EPA METHOD 524.2, Measurements of Purgeable Organic Compounds in Water by Capillary Column Gas Chromatography – Mass Spectrometry*, rev. 4.1, US EPA, Cincinnati (1995).
21. J. A. Glaser, D. L. Foerst, G. D. McKee, S. A. Quave and W. L. Budde, *Environ. Sci. Technol.*, **15**, 1426–1435 (1981).

Index

Quantitative Applications of Mass Spectrometry I. Lavagnini, F. Magno, R. Seraglia and P. Traldi